Fundamentals of the Intelligent Hospital

Over the last decade, the changing healthcare environment has driven hospitals to critically evaluate and optimize their operations to enhance patient treatment and care prompting the emergence of the intelligent health system. Unlike traditional health systems, intelligent health systems are emerging as entities that leverage data, AI, the cloud, and other digital tools to create strategic advantages and better outcomes. While all health systems claim to be doing this, there is a different approach used by intelligent health systems. Both models recognize the inherent power of using data and AI to improve the delivery of health services, but intelligent health systems take this to the next level. They are using the intelligent health revolution to rethink the entire delivery model. Their focus is on leveraging AI and other technologies to efficiently provide health and medical services across all touchpoints, experiences, and channels.

To deal with the complexity of patient care workflows and enhance patient diagnosis, treatment, care, safety and satisfaction, home health, and telemedicine, the design of Intelligent Hospitals has focused on the integration of diverse technologies to provide a seamless exchange of information. This book outlines the technologies and clinical applications which constitute the fundamentals of the Intelligent Hospital and the technologies that support patient care and health management across a spectrum of healthcare environments, the home, remote offices and treatment facilities, and the hospital itself.

The book introduces the changing environment of care resulting in new distributions of patients across a broad spectrum of the patient acuity and care environments, including the Intelligent Hospital; supporting treatment areas, such as the OR, Radiation Therapy, interventional radiology, and patient care areas, such as the ICU, ambulatory/telemetry; and supporting functionality, including infection control, laboratory medicine, pathology, biomedical engineering, Informatics, and Information Technology; the iHome is networked to dynamically interact with the technologies and application in a variety of care settings.

Each chapter describes the fundamentals of integrating diverse technologies and clinical systems to create a seamless environment enabling data to be shared across the complete spectrum of care environment, starting from the ambulance, through the emergency room, diagnostics, treatment, and recovery in the hospital, to the home environment. The book highlights and details specific topics, including medical devices and applications, RFID, network and wireless infrastructure, interoperability and integrations, artificial intelligence, and much more, with a focus on connectivity, integrations, and architecture to create a seamless environment.

Intelligent Health Series

Closing the Care Gap with Wearable Devices

Innovating Healthcare with Wearable Patient Monitoring
Mike Davis, Michael Kirwan, Walt Maclay and Harry P. Pappas
2023

The Rise of the Intelligent Health System
Paul Frisch & Harry Pappas
2024

Fundamentals of the Intelligent Hospital
Adapting Diverse Enabling Technologies to Transform Healthcare Delivery
2025

Fundamentals of the Intelligent Hospital

Adapting Diverse Enabling Technologies to Transform Healthcare Delivery

Paul Frisch, PhD

Routledge
Taylor & Francis Group
A PRODUCTIVITY PRESS BOOK

Designed cover image: Shutterstock

First published 2026
by Routledge
605 Third Avenue, New York, NY 10158
and by Routledge

4 Park Square, Milton Park, Abingdon, Oxon, OX14 4RN

Routledge is an imprint of the Taylor & Francis Group, an informa business

ISBN: 9781032539713 (hbk)
ISBN: 9781032539706 (pbk)
ISBN: 9781003414568 (ebk)

DOI: 10.4324/9781003414568

Typeset in ITC Garamond Std
by KnowledgeWorks Global Ltd.

Dedication

This book is dedicated to my wife, Susan, and our children and their spouses, Paul & Katie, Lauren & Tom, and Daniel & Gabby, and our grandchildren, Reagan, Mia, Austin, Zoey, Landon, Collins, and Dylan, who provided me the inspiration to contribute to the work highlighted in this book. Looking toward the future, my wish is that the continued advancement of healthcare technologies will have a significant positive impact on the health, wellness, and longevity of my family and all families across the world.

Paul Frisch, PhD

Contents

Preface

Introduction

Over the course of the last decade, hospital-based healthcare systems have continued to experience an increase in patient acuity levels, driving an increase in the number of critical care monitored beds, emergency room visits, and post-surgical cases. With the advent of diminishing family practices, many patients use hospital emergency departments as their primary method of general healthcare, significantly impacting patient volumes, flow, and throughput. In 2019, the Coronavirus (COVID-19 pandemic) stressed hospitals beyond their capacity with large numbers of patients presenting to the Emergency Departments which directly resulted in an increasing percentage of admissions. This in turn stretched and exceeded the medical resources, including patient beds, medical equipment, supplies, as well as the medical staff necessary to manage these patients effectively. These, as well as financial constraints and factors, have driven an increased demand for an enhanced healthcare institution that focuses on an advanced seamless medical environment to deliver advanced patient treatment and care optimizing patient outcomes, safety, and satisfaction and optimizing operational and business workflows and operations.

To compensate for the changing healthcare environment, the evolution of the Intelligent Hospital has capitalized on new evolving and enabling technologies integrating and communicating with medical device and clinical software applications to create a seamless continuum of care to improved patient outcomes, enhances patient care, safety, and satisfaction through optimized clinical and business workflows and supply management increasing the effectiveness and reducing the cost of operations. The book is divided into four main sections, including supporting appendices to provide additional detail and content:

- Adapting and Integrating Diverse Enabling Technologies
- Impacts on the Patient Care Workflow
- Developing Patient Specific Solution to Enhance Patient Care Delivery
- Management of These Complex Systems and the Patient Care Environment

The first section of the book examines a variety of diverse enabling technologies which, when integrated with advanced medical devices, systems, and clinical applications, contribute to the development of a seamless patient care environment providing standardization and new capabilities for patient centric operations, visualization, and targeted communications. These integrations have evolved into complex medical systems enabling critical information directly impacting patient outcome, care, safety, and satisfaction to be rapidly disseminated to the point of care optimizing clinical and business processes and workflows. This section highlights a variety of integrations which target and enable a dynamic interactive seamless data environment while supporting functionality to dynamically monitor patients and validate processes and workflows to reduce errors and enhance safety.

Section 2 provides detailed examples of how these enabling technologies directly impact patient care workflows and scenarios providing comprehensive examples of realistic patient care process in a variety of care environments, the ED, ICU, Stepdown/Ambulatory Care, OR, as well as several specialty areas such as the Cath Laboratory and a centralized monitoring war room for supporting patient care seamlessly throughout the patient care scenario, including telemedicine and home care.

Section 3 takes a deeper dive into specific new technologies providing advanced capabilities and functionality for targeted patient specific treatment solutions and guides and modalities. This includes advanced 3D imaging capabilities which, when coupled with 3D printing and AR/XR visualizations, enable a new dimension in patient diagnostics and treatment planning. This section examines how 3D technologies create anatomically accurate models, custom surgical instruments enabling advanced visualizations to target treatment modalities to maximize patient outcome and provide enhanced safety.

Section 4 examines the support modalities needed to maintain and manage these technologies to ensure operational integrity and operational consistency and to provide a dynamic and safe patient care environment. Fundamentally, medical devices/systems are no longer stand-alone devices operating independently. These systems are integrated to create new complex systems, integrating multiple medical devices with each other, with clinical software applications and enabling technologies all operating together rapidly passing critical information on the Hospital's wired and wireless network infrastructure. The support of these complex systems requires a new and expanded dimension of support including many operational synergies within the hospital. These new synergies require an advanced support strategy to including Biomedical Engineering, IT, Facilities, and the Clinical Departments to provide the dynamic support strategy for these complex systems. These strategies also require a new type of support environment including a simulation facility dynamically simulating the hospital's patient care environment in terms of devices, clinical systems, and supporting technologies. This simulation environment provides isolated capabilities to accurately and dynamically test systems for new upgrades, including hardware and software, software patches, technology enhancements, and new integrations in a stable offline environment to minimize unforeseen interactions and operation impacts which can impact operational integrity and patient safety. Clinical staff can use this simulation space to develop new workflows to optimize patient care and provide education and training. Cybersecurity has also added a new dimension for the testing and implementation of controls to minimize vulnerability to threats impacting functionality and patient safety.

It is the author's hope that this book will provide insight into the potential advantages of this continuously evolving Intelligent Hospital environment and has highlighted the fundamental considerations required to manage and maintain this advanced patient care environment.

Paul H. Frisch, PhD

Section 1

Enabling Hospital Intelligence

Introduction

The Intelligent Hospital is defined as an enhanced and adaptive patient care environment exploiting new and evolving technologies to establish a standardized and seamless continuum of care. The Intelligent Hospital concept focuses on the development and integration of a variety of diverse and enabling technologies with advanced medical devices and clinical applications/systems enhancing the effectiveness of clinical workflows and processes. These enhanced workflows optimize the clinical and business processes and operations and directly impact improved patient outcomes, enhancing patient care, safety, and satisfaction. This book addresses a broad set of topics which highlight the technologies, integrations, processes, workflows, and their combined impact on patient care and management, safety and satisfaction, and the potential of optimizing operating processes and costs.

The continuum of care (Figure 1.1) can be defined in terms of the patient experience of the care processes as they transition through the many diverse hospital scenarios and environments, including triage, diagnosis, treatment, recovery, and follow-up. As an example, the patient experience can start and include a combination of scenarios such as

1. Ambulance and Emergency Department visit.
2. Triage and diagnostics can include laboratory tests, and imaging, which may include a variety of Radiological imaging modalities including CT, MR, PET, or ultrasound.

DOI: 10.4324/9781003414568-1

3. Treatment, which can be invasive, such as surgery or cardiac catheterization, or longer-term therapies, including radiation therapy or chemotherapy.
4. Acute care, which can be experienced in the critical care areas (ICU, step down); and recovery and general patient care.
5. Follow-up care can occur in the hospital or in regional ambulatory care areas or in medical office visits.
6. Patient management and long-term monitoring, which can occur utilizing remote telehealth or telemedicine applications enabled through in-home technologies (i-Home).

The evolution of the Intelligent Hospital is a continuous process focused on utilizing a combination of new and existing medical systems, clinical applications, and technologies simultaneously exploiting new and evolving technologies to enhance the clinical, operational processes, and business workflows, providing a seamless data environment to enhance and optimize the patient care process as well as operational business processes as the patient transition through the different care environments and facilities associated with their care.

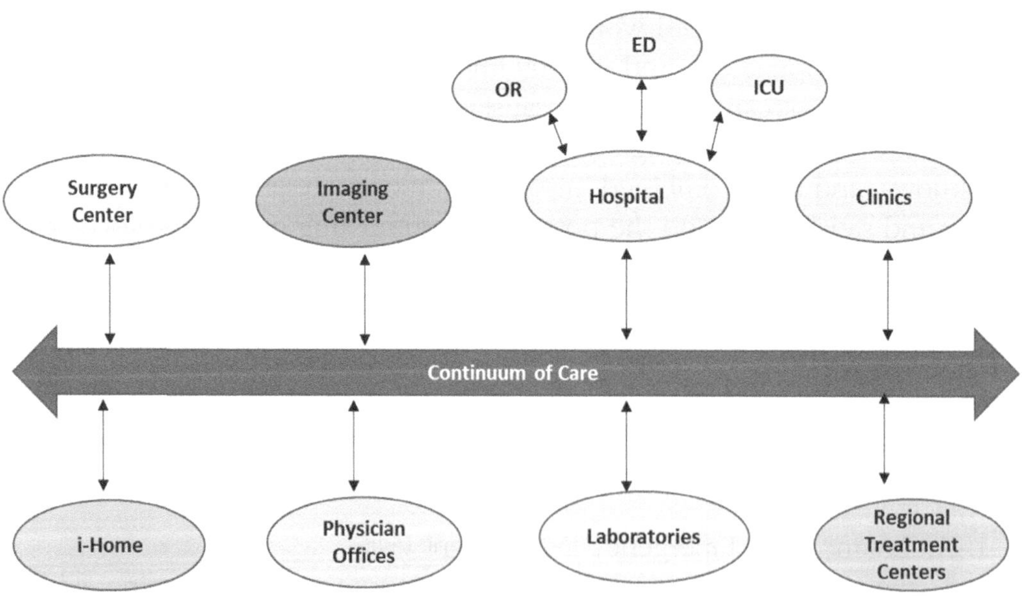

Figure 1.1 The Intelligent Health System continuum of care.

Drivers for Enhanced Hospital Intelligence

Over the course of the last two decades, hospital-based healthcare systems have continued to experience an increase in patient acuity levels, driving an increase in the number of critical care monitored beds, emergency room visits, and post-surgical cases [1–3]. With the advent of diminishing family practices, many patients use hospital emergency departments as their primary method of general healthcare, significantly impacting patient volumes, flow, and throughput. In 2019 the Coronavirus (COVID-19 pandemic) stressed hospitals beyond their capacity with large numbers of patients presenting to the Emergency Departments which directly resulted in an increasing percentage of admissions. This, in turn, stretched and exceeded the hospital's medical resources, including patient beds, medical equipment, supplies, as well as the medical staff necessary to manage these patients effectively. As a result of the pandemic, patients deliberately delayed any non-critical procedures to limit their potential risk of exposure in hospitals. In response, hospitals were forced to exploit telemedicine or telehealth applications driving a growing demand for home health and monitoring applications [4].

The combination of these factors has been identified and associated as contributors to the increasing number of errors, including medication errors, device and systems operating errors, and errors associated with patient identification, resulting in an increased risk to the patient as outlined below.

Increasing Medical Errors

- Medication Errors [5–7]
- Alarm Management and Fatigue
- Device Operating Errors
- Identification Errors

As a note of caution based on various sources of this information, the statistical data associated with these events can represent a broad range of values which are dependent on the criteria used to determine the statistics and should not be used at face value. However, these trends are so significant and have raised extremely high levels of concern within regulatory agencies and hospital accreditation organizations, such as the Centers for Medicare and Medicaid Services (CMS) and the Joint Commission (TJC). In response, TJC has established guidelines (National Patient Safety

Goals) to address these patient safety concerns [8]. For more than a decade, TJC's patient safety goals have targeted medication-related errors, the need for improved identification and association processes, improved validation and verification processes, and enhanced communications capabilities between staff-to-staff and patient-to-staff. The technologies which positively impact these safety concerns have become some of the key drivers for the evolution of the Intelligent Hospital concept.

To compensate for the changing healthcare environment, hospitals found it necessary to critically evaluate, improve, and optimize their clinical and business processes to enhance patient outcomes, diagnostics, treatment, and care and maximize patient safety while optimizing resource utilization and patient throughput. These objectives have driven an increased interest and reliance on new and evolving enabling technologies which can be used to directly impact, enhance, and optimize the clinical and patient care processes. The Intelligent Hospital has exploited these technologies to provide connectivity, advanced communications, and visualization supporting a seamless exchange of information via a high-speed, high-bandwidth wired and wireless infrastructure across the entire continuum of care. This complex interconnectivity has identified the need for accurate time synchronization between systems as well as the need for patient centric identification and association of medical information to support the clinical and patient care and management processes. This dictated the need for establishing a standard methodology used to identify patients, staff, clinical supplies, pharmaceuticals, care locations, and medical devices.

Objectives of Intelligent Hospital

The primary objective of the Intelligent Hospital is to establish a seamless and standardized patient care environment across the entire hospital environment of care or continuum of care. The Intelligent Hospital focuses on the development, integration, and deployment of a combination of diverse enabling technologies with advanced medical devices, clinical applications, analytics, and commercial IoT devices to seamlessly deliver accurate and consistent actionable information across the continuum of care.

As illustrated in Figure 1.2, medical devices and clinical applications share information via the hospital's network and wireless infrastructure. All data is associated with specific patients, which is then associated with assigned clinical care providers and resides in shared data archives, which enables data to be aggregated, processed, filtered, validated, and delivered to any location or IoT device within the environment of care. Critical data, including physiological parameters and alarm/alerts, can be prioritized and delivered to decision support and intelligent systems or directly to clinical staff at the point of care to provide rapid response and an enhanced level of patient care. Aggregated patient specific information can also be transmitted, displayed, and visualized on dashboards, in a variety of formats and locations, including a central war room or mission control area to improve oversight, operational efficiency, and effectiveness and optimize resource utilization and management.

The Patient Care Process

Understanding the patient care process and flow of the patient throughout the environment of care provides the basic groundwork for establishing the technical requirements of the Intelligent Hospital. It defines the actionable data and information flow necessary for establishing continuity

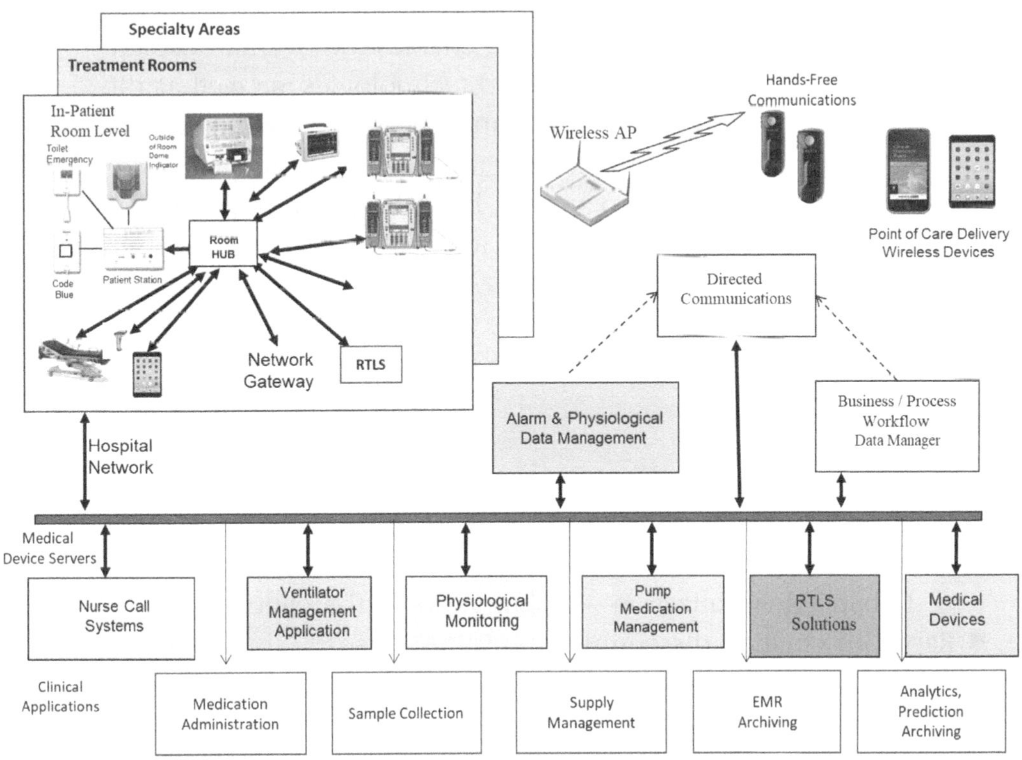

Figure 1.2 Overview of the Intelligent Hospital.

in the care process and workflows throughout the hospital. The patient process frequently starts either in the physician's office or in an emergency scenario, such as the ambulance, or in the Emergency Department where initial triage and immediate care are provided. In this phase, patient demographics and medical history are verified or identified and loaded into the patient medical record. Unique patient identification is critical to establish the relationship to the clinical patient information as well as a link to historical medical information which may be available and relevant. Complex diagnostic decisions are supported by physiological data, specialized imaging modalities, and clinical laboratory results. Based on the diagnosis, specific treatment plans are developed which can include further testing and evaluation, surgical procedures, or a variety of specialized treatments such as radiation therapy, dialysis, chemotherapy, and physical therapy. Post treatment patient monitoring and recovery continues in the acute care ICU, step down, ambulatory care area, or general care and observation areas and can also include continued monitoring of patient status post discharge in the home.

The design, development, and implementation of the Intelligent Hospital are influenced by many factors. Few health systems have the opportunity to design a hospital from scratch, which enables the overall hospital architecture, infrastructure, and required technologies, as well as the supporting budgets to be coordinated and developed simultaneously. Most health systems are forced to implement the Intelligent Hospital incrementally in phases over many years of planning, budgeting, and implementation which also dictates that the design and implementation accommodate and adapt to existing processes and incorporate legacy technologies and systems.

The overarching objectives of the Intelligent Hospital can be summarized as follows:

- Increased data accuracy and availability
  - Electronic medical records
  - Shared medical device interconnectivity and integration
  - Enhanced visualization – global views of the patient/unit
- Rapid information dissemination and response
  - Physiological data
  - Alarm event management
  - Advanced communication platforms (PDA, iphone, etc.)
  - Location content

- Enhanced error checking and validation
  - Identification and association
  - Location content
  - Process validation
- Optimization of workflow and resource management

The vision of Intelligent Hospital is focused on establishing a methodology to seamlessly collect, share, and distribute accurate and validated information across the entire spectrum of patient care. To establish this seamless patient centric environment, the Intelligent Hospital design requires and utilizes many diverse technologies linked through the robust networked infrastructure capable of supporting real-time data exchange via a high-speed, high-bandwidth wired and wireless infrastructure. Integration between medical devices, supporting sensors, IoT technologies, analytics, and clinical applications enables a broad spectrum of patient centric solutions to be developed and deployed. This interconnectivity produces large volumes of validated and actionable data which can be aggregated, filtered, processed, archived, and rapidly shared or disseminated to any location within the hospital including a combination of a central war room, dashboards, and directly to a handheld communications device at the point of care.

As summarized below, the fundamental component of the Intelligent Hospital includes medical devices, enabling technologies, analytics, and clinical applications which create the many use cases to produce a high-performance and dynamic healthcare environment.

- Advanced medical devices
  - Transducers and sensors
- Networked and wireless infrastructure
- Time synchronization
- Patient centric identification
- Centralized or shared data acquisition – connectivity of devices
  - Physiological and Alarm Data Management
- Analytics and decision support
- Location-based systems, RFID, RTLS, BLE (Bluetooth Low Energy)
  - Device tracking
  - Workflow monitoring
- Medication administration
- Sample collection

- Video and imaging
- Global/Institutional overview and visualization of rooms/units
- Enhanced communications
- Point of care information delivery
- Inventory management
- Supply chain
- Technology support and management

Components of the Intelligent Health System

Though this above list primarily highlights electromechanical and IT based innovations and enabling technologies, it is important to note that other advancements and technologies also directly impact the design of the Intelligent Hospital. These include materials science, chemical, and biological developments which directly impact and result in advanced construction techniques and are designed to improve patient safety and satisfaction which are also critical factors that contribute to the overall Intelligent Health System design and implementation.

The following highlight several examples, which can have a significant impact on patient care, safety, and the overall patient satisfaction experienced based on their interaction within the surrounding environment of care.

- Antimicrobial copper
- Patient privacy e-Glass
- Personal patient/staff sanitizer solutions

Antimicrobial Copper Surfaces

Antimicrobial copper can be used to replace stainless steel on high-contact infection-prone surfaces, including handrails, counter tops, and door handles, as illustrated in Figure 1.3. Copper surfaces have a proven capability to minimize surface bacterial or microbial loads reducing the potential of infection transmission.

Patient Privacy e-Glass

In the Intelligent Hospital, patient privacy e-Glass deployed within the care environment enables rapid privacy often critical during emergency situations, treatment, as well as during routine patient care. Privacy is assured using

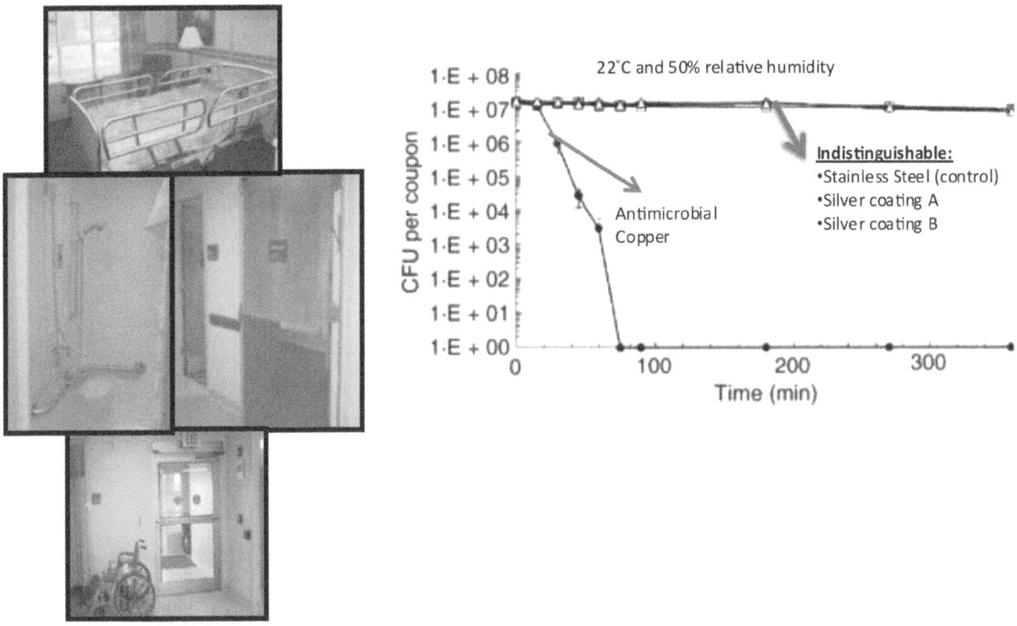

Figure 1.3 Applications of antimicrobial copper in the environment of care.

electronic switchable privacy glass, sometimes known as e-Glass or LC Privacy glass. The glass panels are filled with electrically stimulated gas. Once an electric charge is applied, the gas turns from clear to opaque. This glass can easily be cleaned and replaces the ubiquitous curtains which are easily stained and subject to bacterial infection. The e-Glass can be installed in swinging doors, as well as in sliding breakaway doors, and controlled with a simple switch located in the room or bedside.

Lighting Solutions

Lighting solutions can have a significant impact and perception of the care environment. Optical solutions based on nanotechnology can artificially reproduce the natural light and visual appearance of the sun and sky. As illustrated in Figure 1.4, a typical use is demonstrated within a patient room. The lighting solution produced a blue sky and shining sun creating a positive impact on the clinical space. In many care environments, which are typically dark with windows, such as radiation treatment areas and operating rooms, the potentially claustrophobic space can introduce stress which negatively impacts the patient. The representation of sky and sunlight is therapeutic increasing patient humor and satisfaction.

Figure 1.4 Artificial lighting solution.

Innovative Hand Hygiene Solutions

According to the U.S. Centers for Disease Control, hospital-acquired infections (HAIs) cause 1.7 million infections, with 99,000 deaths each year [9]. As illustrated in Figure 1.5, the use of a point of care hand sanitizing solution can significantly increase compliance. The device is a single hand activated device and helps to decrease HAI rates up to 40% beyond using only wall mounted units and delivers a premium 72% ethyl alcohol gel sanitizer through a patented roller-ball system, effectively killing 99.9% of germs. Worn as recommended, the device is ideally positioned between the contaminated hand and the patient, object, device, or procedure at risk of being contaminated and if used properly increases hand hygiene compliance.

Design and Development of the Intelligent Hospital

As illustrated in Figure 1.6, the operations of the Intelligent Hospital revolve around many clinical and operational components distributed throughout the institution functioning to provide emergency care, diagnostics,

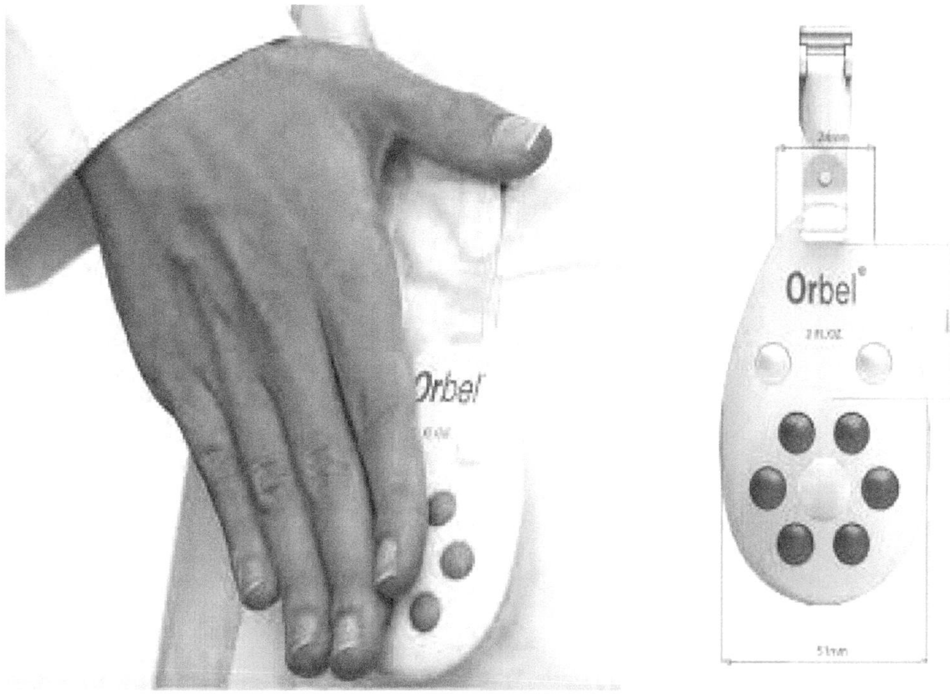

Figure 1.5 Point of care hand hygiene solution.

treatment, and patient management. These functions are performed by many clinical/medical specialties such as Cardiology, Radiology, Surgery, Radiation Oncology, Labor and Delivery, etc. These clinical processes are further supported by many operational groups, as outlined below, delivering support to the clinical processes and patient management to achieve the overall objectives of patient care and management in the Intelligent Hospital.

■ Supply Chain	Supply Management and Distribution
■ Infection Control	Managing Cleaning and Sterilization Process
■ Clinical Engineering	Ensure Medical Equipment Integrity
■ Information Technologies	Ensure Cybersecurity and Operation of Infrastructure and Devices
■ Facilities Management	Ensure a Safe Patient Care Environment
■ Food and Room Services	Managing the Patient Environment and Food Delivery

As the patient transitions through the care process, the patient is exposed and moves through a variety of care environments starting from an

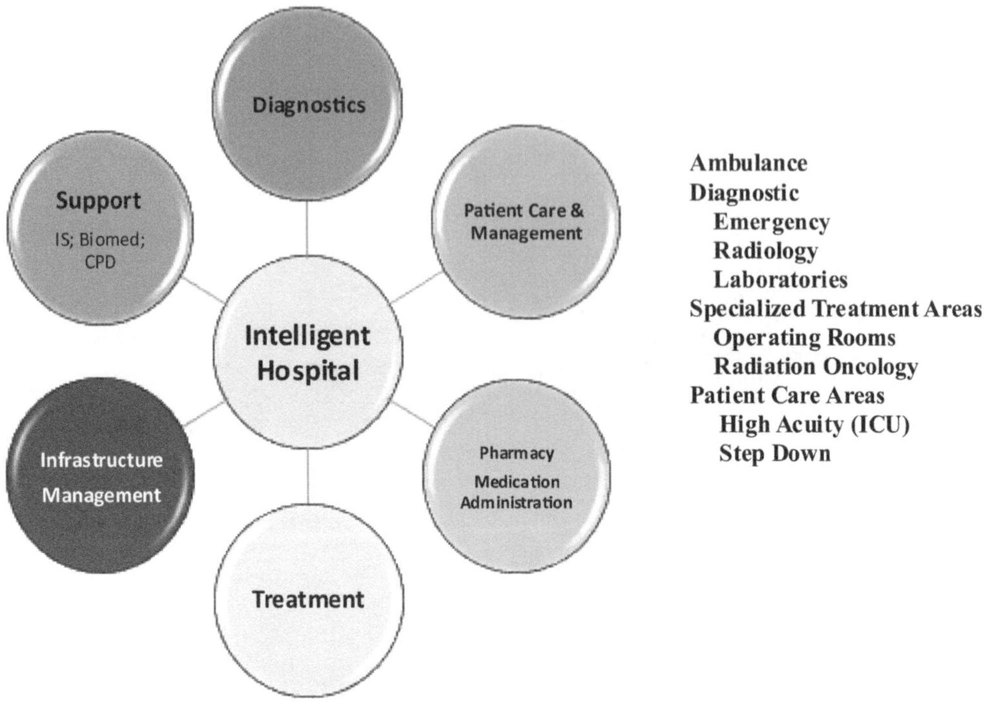

Figure 1.6 Hospital operations and processes.

emergency response and triage environment (Figure 1.7), moving through diagnostic phases, which include laboratory and imaging to a treatment which can include surgical (OR) options (Figure 1.8) to recovery including varying degrees of patient care management which includes the ICU, step down, or general care areas (Figure 1.9). The specific medical devices, applications, technologies, and workflows can vary significantly as a function of each care environment.

Standardization of Technologies within the Patient Care Environment

A critical design consideration is the standardization of clinical data, dataflow, and workflows, which in turn drives the selection of medical devices, clinical applications, enabling technologies, identification methodologies, and patient association with the clinical processes. The combination of technologies can significantly vary between the different care areas based on the patient diagnosis, acuity, and the required clinical functionality and workflows.

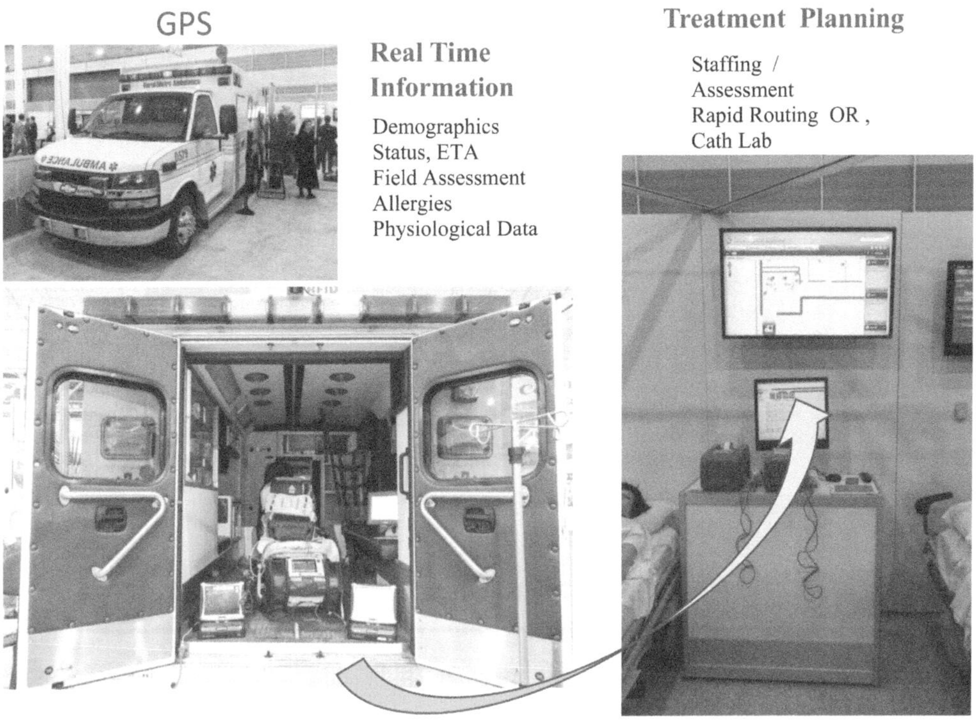

Figure 1.7 Ambulance emergency response.

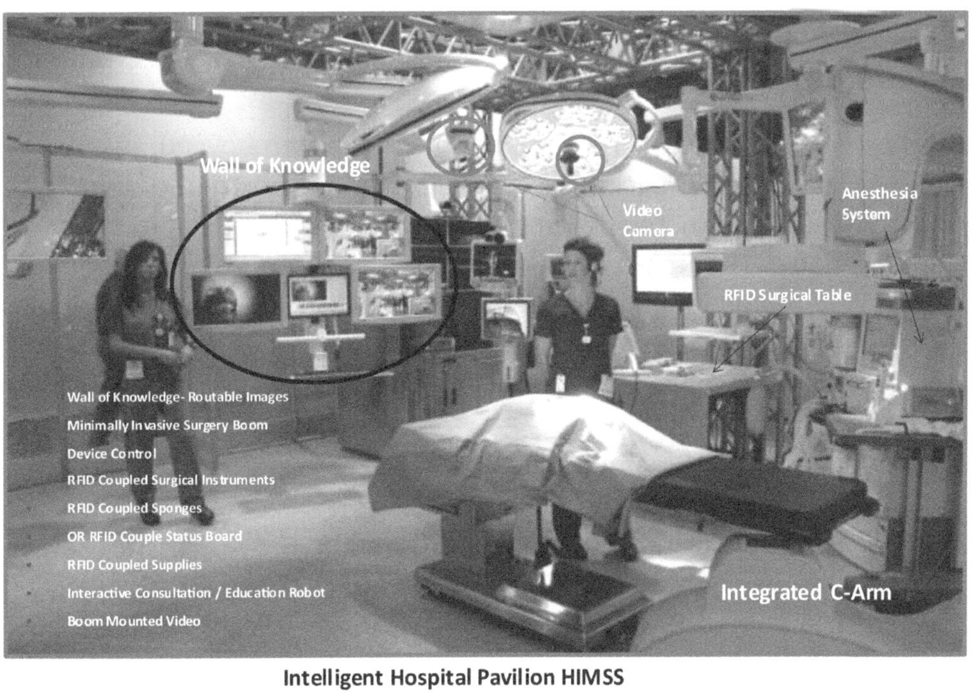

Figure 1.8 Integrated hybrid OR.

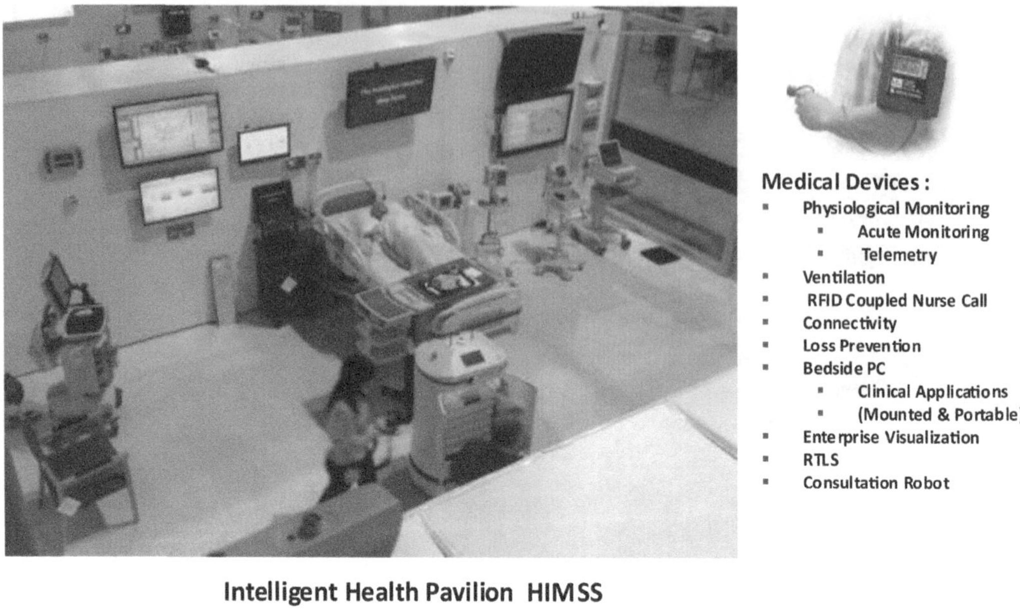

Figure 1.9 ICU/patient care room.

Clinical functionality is commonly classified into several care categories, where device functionality and utilization are defined as a function of patient care modality (diagnostics, treatment, and management).

Institutional Infrastructure

Ideally, the specific systems that are common across the overall institution establish a standardized backbone across the entire spectrum of the hospital or healthcare network. This backbone is fundamental to enable communications, data transfer, and establish a seamless data environment across the entire continuum of care. These would include a variety of shared network technologies, common data and patient archives, devices, clinical applications, and enabling technologies as outlined.

- Network and wireless infrastructure
- Electronic medical record
- Medication physician order and administration
- Communications and IoT devices
- Medical systems and devices
- Supply chain management
- Real-time location technologies

Specialized Systems and Devices

All hospitals deploy many specialized systems designed for targeted functions which are unique to a patient's treatment and care processes. These are typically specialized high-end devices which are selected as the best of breed systems unique to specific diagnostic functions, treatments, and care scenarios. It is critical that these devices also support connectivity to the institutional infrastructure and maintain a seamless data architecture exchange protocol.

- Imaging and radiological systems
- Laboratory systems
- Pharmacy systems
- Catheterization lab
- Anesthesia system
- Surgical robots
- Laser systems
- Linear accelerators
- Brachytherapy

Standardized Institutional Patient Care and Management Systems and Solutions

These examples typically include medical devices and clinical systems that provide functionality applicable to a broad base of patients focused on patient management, monitoring, and care. These institutional solutions, including medical devices, clinical applications, and enabling technologies, target use cases and workflows across all patient care areas in the hospital.

One of the fundamental principles and objectives in the design and implementation of the Intelligent Hospital concept is the creation of a seamless data environment providing a validated and patient centric set of information across all the care environments. To create a seamless data environment, standardization is vital and must occur at many levels. Within many hospitals, standardization is frequently poorly defined or inconsistent because of implementing unit specific isolated legacy solutions over many years. Standardization is a key factor impacting the integration and sharing of data between medical devices, clinical applications, enabling technologies, and data analytics. The following highlights some of the considerations for standardization goals within the Intelligent Hospital.

Potential Standardization Targets within the Intelligent Hospital

- **Medical Devices**
  - Beds/Nurse call
  - Physiological monitoring
  - Ventilators
  - Infusion pumps systems
- **Clinical Applications**
  - Patient admissions and discharge
  - Electronic medical record
  - PACS
  - Medical order entry and delivery
  - Supply chain management
  - Medical device inventory system
- **Enabling Technologies**
  - Connectivity solutions
  - IoT devices (communication/data delivery)
  - Identification and Real-Time Location Technologies (RFID/RTLS/BLE)
  - Communication devices and IoT devices
- **Identification Nomenclature**
  - Patient identification
  - Medical device ID
  - Pharmaceutical ID
  - Supply chain ID
    - Unit/Care area location ID
    - Bed/Patient unit level location

Standardization identifies and drives the requirements for the infrastructure enabling data to be shared and presented seamlessly across the institution. It minimizes the operational costs associated with the development of integrations between systems, reduces the need for data conversion and formatting, minimizes staff training and reduces the support and service, and minimizes the need for continued education.

The level of standardization will vary between institutions and will be largely based on the existing legacy systems and the number of required integrations necessary to achieve the expected level of shared information and optimization of the workflows. It is important to note that standardization is a continuous process, and needs occur at many points during the development and implementation of the Intelligent Hospital. However, a successful standardization process significantly impacts the success of a robust seamless environment enabling connectivity, data transfer, and visualization.

Fundamental Design Components of the Intelligent Hospital

Complex Medical Systems – Advanced Medical Devices, Transducers, and Sensors

Recent innovations have enabled extensive advancements in medical devices, transducers, and sensors enhancing patient care across all modalities of the care process from triage, diagnosis, treatment, care delivery, and recovery. Many Biomedical/Clinical Engineering books, publications, and manufacturer literature are available to provide comprehensive details on specific medical device operational capability and functionality [10–13]. Since the specific complement of the medical devices used at healthcare institutions can vary significantly, the specific product device specifications will not be repeated in this book. Please refer to the broad base of literature available on today's advanced medical systems, devices, transducers, and sensors.

It is very important to note that within today's Intelligent Hospital, medical devices are no longer stand-alone devices operating asynchronously but are now connected together and integrated with clinical applications, analytics, and IoT devices, creating complex medical data systems providing decision support, data archival, and data visualization and delivery. This complex interconnectivity has clearly defined the need for a robust high-speed infrastructure/networking architecture, accurate time synchronization, and standardized unique identification to provide the operational functionality required within the Intelligent Hospital. These complex systems and functions do require new and enhanced support management strategies to validate device functionality and operation, which will be discussed further in subsequent chapters of the book.

The following examples highlight several complex data systems common to the Intelligent Hospital.

Example #1: Physiological Monitoring System

As illustrated in Figure 1.10, the complex system architecture is based on a bedside physiological monitor capable of operating asynchronously, measuring and acquiring patient physiological parameters including heart rate, ECG, blood pressure, end-tidal CO_2, pulse oximetry, temperature, etc., and displaying this real-time information at the bedside monitor. On a unit level, multiple monitors connect to a central station where aggregate data from multiple patients is displayed. These monitors are typically interconnected via a dedicated network or a standard TCIP network and are also capable of wirelessly (telemetry) sharing data between these systems. These networked monitors pass the physiological parameters to one or multiple central stations or dashboards within the unit. All the central stations can be further networked creating a dedicated hospital network and then via the hospital network backbone be connected to other sites and multiple clinical applications. In this example, the physiological monitoring network interfaces and passes data to remote monitoring systems for visualization, long-term data archival, interfaces to ventilators, expanded archival to the patient medical record (EMR), and multiple communication applications to deliver information to a variety of locations including the point of care. This complex medical data system enables data to be seamlessly available across the institution, to provide enhanced patient monitoring care and alerts.

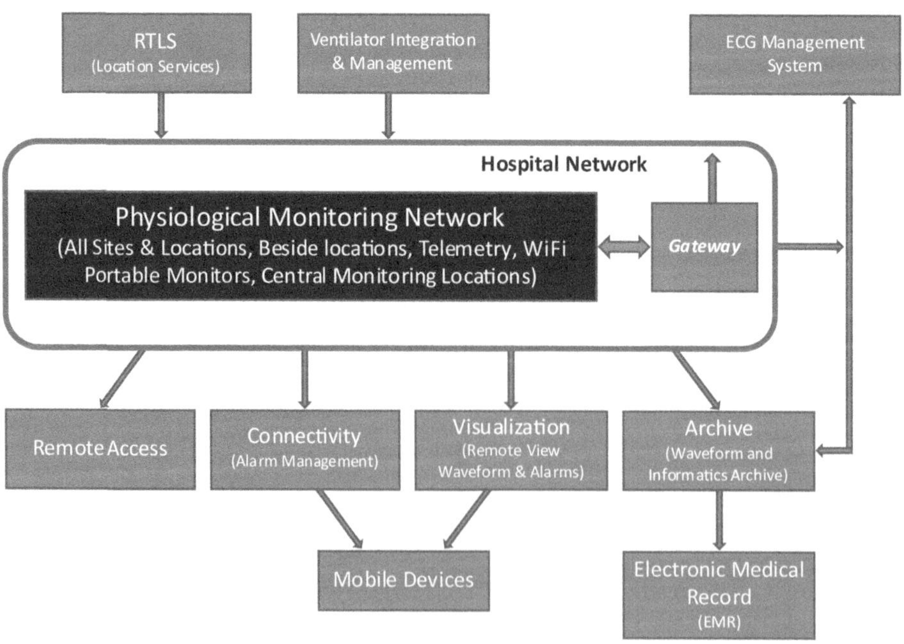

Figure 1.10 Advanced integrated complex medical system.

Network and Wireless Infrastructure

As previously discussed, the fundamental linking technology component to provide a seamless shared data environment is the underlying infrastructure providing a high-performance network consisting of a robust high-speed network architecture including a blend of multi-carrier cellular, Wi-Fi services, and dedicated wireless frequencies. Together, these support the specific functionality required for complex medical systems, such as physiological monitoring/telemetry, to be coupled with active and passive RFID, RTLS solutions, BLE, and advanced communications to create a broad base of use cases enabling a variety of optimized workflow functionality. This critical network infrastructure is a fundamental component of the Intelligent Hospital architecture and the key enabler for many of the new and enabling technologies and applications providing the medium for real-time high-speed information exchange throughout a hospital and the overall hospital network.

This network architecture establishes the connectivity and provides enterprise-level operational intelligence and the opportunity for real-time data archival, visualization, and clinical response which results in a positive patient clinical impact. All clinicians and medical staff can now utilize mobile handheld IoT devices enabling them to review, communicate, and pass forward critical information, such as device alarms, imaging, and clinical lab data for review and consultation. Within the Intelligent Hospital, these devices are commonly integrated into the hospital phone system, additionally enabling incoming and outgoing calls directly from the point of care, increasing staff presence at the point of care (bedside), and further enhancing patient care and satisfaction. Medical devices, such as physiological monitoring, are commonly networked to clinical applications, such as the EMR, to enable collection and archiving of a patient's critical parameters and vitals as they transition through the various care environments. This capability establishes a continuous and retrievable record of the state and status of the patient at any point in the care process enabling changes in state patient and trends to be closely monitored, analyzed, and managed.

Time Synchronization

To accurately compare and analyze data from multiple devices, systems, and other sources, the Intelligent Hospital requires that all data parameters and information be time synchronized throughout the entire care environment. This ensures that treatment and care decisions are made effectively based on accurate patient data and clinical documentation. Time synchronization is an effective way to ensure operational efficiency within the healthcare environment, establishing increased patient safety and standard quality of care. With accurate, synchronized time throughout the hospital, staff can be sure treatments and medications are administered at the appropriate scheduled times. The system communicates over a standard network (hardwired or Wi-Fi) to keep all clocks and timers in sync and provide actionable data that is time synchronized and accurate.

Environmental Sensing

The Intelligent Hospital infrastructure ideally incorporates real-time continuous environmental surveillance monitoring of key environmental parameters such as temperature, humidity, airborne particulates, differential room pressure, light and sound, and air quality. Data from these sensors provides analytic information used for infection control and facilities management enhancing patient safety and satisfaction. Real-time alerts are generated anytime a potentially hazardous or undesirable environmental condition is detected, allowing for immediate response and remediation. This monitoring enables critical devices such as freezers to be continuously monitored ensuring regulatory compliance as well as the integrity of critical specimens, laboratory samples, and pharmaceuticals. This configuration enables data to be directly routed to key staff, enabling rapid response, as well as presented in a central war room or review area.

Handheld and IoT Devices

The network and wireless infrastructure not only supports medical devices and clinical applications but also simultaneously supports many commercial products such as dashboards, mobile handheld and wearable devices enabling them to readily communicate, receive, and deliver critical information, including alarms, clinical parameters, or imaging and lab data. These handheld devices are commonly deployed at the point of care but can simultaneously be integrated into the hospital communications or phone systems enabling incoming and outgoing calls and the ability to support real-time remote consultation. Standardizing on the devices used within the health system significantly reduces the number of types of devices used within the environment of care as well as limiting education and training while enhancing patient care by maximizing staff time at the bedside.

Patient Centric Identification and Association

The key to enabling institution-wide seamless environment requires a standardized methodology for identification and association of patients with clinical equipment, care providers, pharmaceuticals, supplies, etc. [12]. This is a key process to define, validate, and ensure data delivery is accurately delivered and directed ensuring patient safety and optimizing operational workflows. This process is required for many clinical and business applications and operations, including medication administration and sample collection, and is necessary for establishing connectivity or identifying medical devices with a patient, ensuring physiological parameter and alarm accurate delivery of information to a war room, dashboard, or point of care device. Operationally, this process is performed and repeated many times over for each patient, using valuable time and introducing a potential for inconsistencies and errors. In most cases, the staff enter patient demographics (name, medical record number, date of birth, etc.) along with care provider identification into each of the medical systems and applications. This information is also commonly entered at a central location, such as the nursing station, which removes the staff from the patient care environment.

Fundamental to ensuring data integrity and delivery accuracy requires an institutional solution establishing a standardized methodology to identify and associate patient-based data. This association methodology defines and establishes the dynamic relationship between staff, patients, clinical systems, medical devices, pharmaceuticals, and supplies. This process is used repeatedly in the patient's treatment and care processes. Patient centric identification, as illustrated in Figure 1.11, enables the use of multiple technologies to be used in parallel, including barcodes, radio frequency identification (RFID, both active and passive), infrared (IR), and BLE to enable the bedside identification and relationship association. This methodology clearly identifies or defines which nurse is providing care to which patient, which devices the patient is connected to, and defines the specific device parameters associated with that patient. It is important to note that standardization for naming of physical space/locations and devices is also critical to ensure accuracy and enable precise location

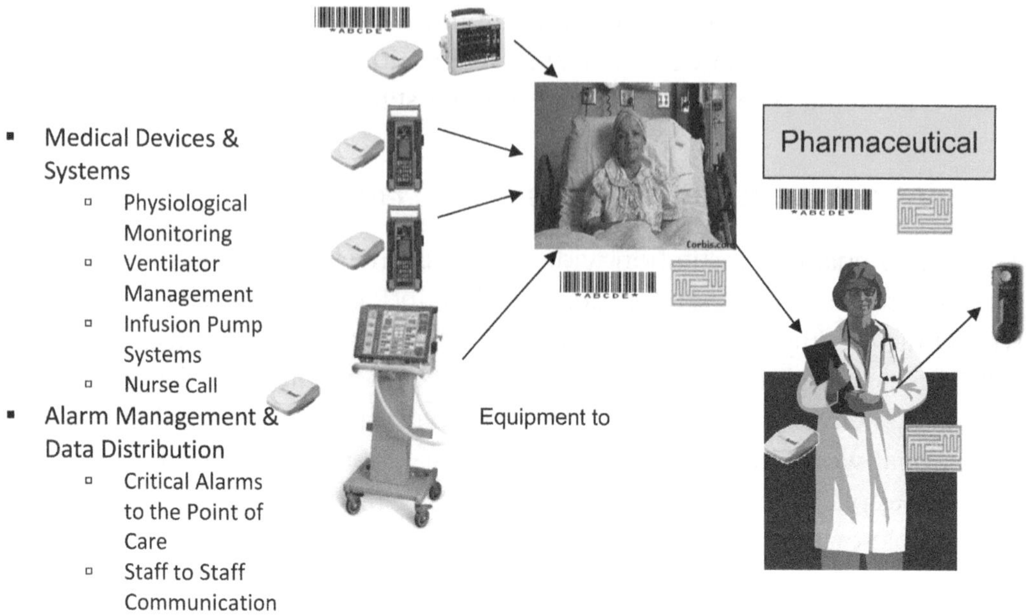

Figure 1.11 Patient centric identification and association.

identification and tracking ensuring that the correct data is delivered to the correct staff person.

In the past, the traditional approach has been to provide association capability at central locations such as nursing stations or staff work areas within the specific device or clinical application. The association process or application integrates with the hospital's ADT (Admission, Discharge, and Transfer) systems for patient information, active directory for staff information, and the biomedical device/asset database for equipment-related information. However, many critical identification processes, including connectivity (defining which device is connected to what patient), medication administration (which medication is being administered to what patient), sample collection (which sample is being taken form what patient), and imaging (image of what patient) typically and optimally, occur at the bedside or the point of care. Ideally, the bedside associations can be stored and distributed to a variety of clinical applications or medical systems reducing the potential of inconsistencies and errors.

As an example, the association process can be performed directly at the bedside using a combination of barcode, RFID, IR, BLE, or RTLS identifiers. These items can be dynamically scanned or identified to be present within a bedside RF, IR, or ultrasound envelope around the bed. To ensure the robustness of this process, a verification or confirmation process of the association can be performed at the bedside and then delivered or forwarded to all applicable applications via an institutional network solution. Once this association is validated and dynamically stored, all devices and applications can access and retrieve any required information defined by this association. Alarm algorithms can retrieve alarms from multiple systems based on the patient association, collate, analyze, and prioritize alarms as a function of the patient. This critical alarm information can then be directed or sent to the associated or assigned care provider, historically trended and archived for review, and delivered to the point of care, central dashboard, or any desired system or archive. This type of association eliminates the need to redundantly define the staff-to-patient relationships typically required and entered into multiple systems and applications (nurse call, monitoring, infusion systems, ventilator systems, etc.). The association process establishes and should support a one-to-many relationship between a single staff member, multiple patients, and many devices, as illustrated in Figure 1.12.

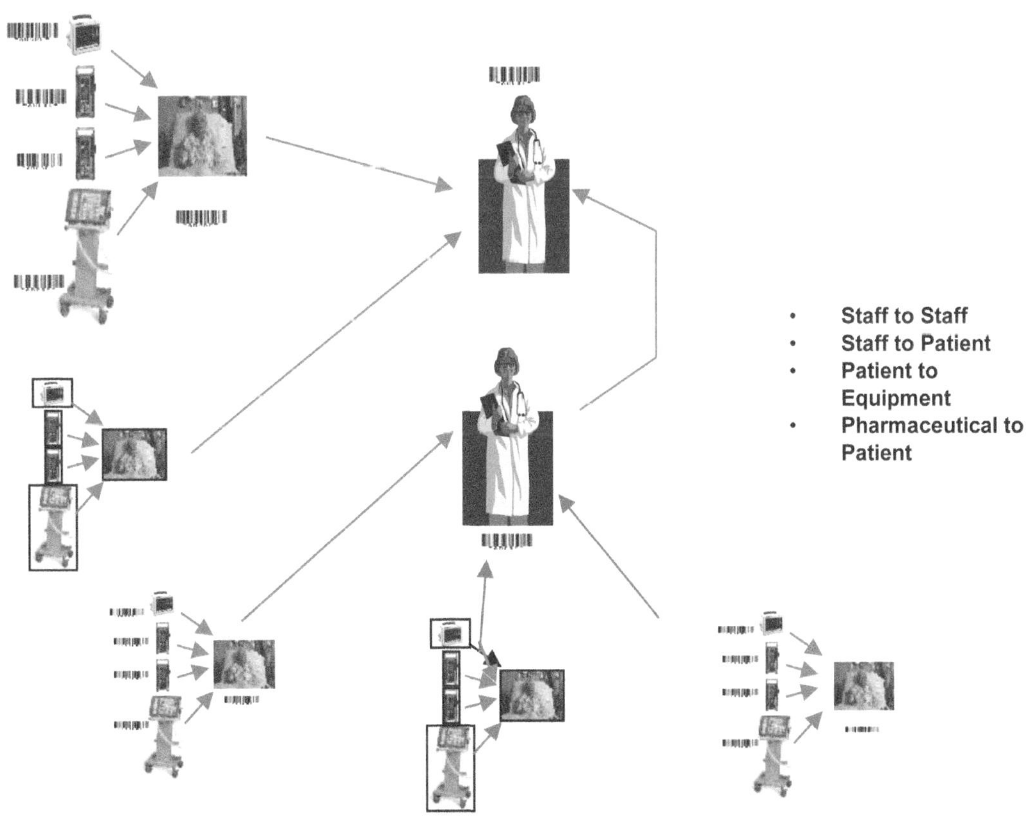

Figure 1.12 Enterprise-wide relationship association.

Connectivity and Integrations

As the acuity of patients increases, patients are typically connected to an increasing number and variety of medical devices including ventilators, infusion pumps, cardiac output devices, etc. It is vital that data from these discrete devices and systems is time synchronized, and patient specific data is aggregated to provide a comprehensive view and interpretation of the patient status and condition. As illustrated in Figure 1.13, the process of acquiring clinical or physiological parameters and alarm information from multiple medical devices and passing or delivering this data to a central processing system or archive defines the connectivity application. Connectivity of medical devices can be achieved via dedicated output connections, i.e., RS232, USB, etc., and/or via standard wireless or wired network protocols. Wireless networks can include Wi-Fi, Wireless Medical Telemetry Services, Bluetooth, or dedicated frequencies to transfer data and support the connectivity process. For medical devices, the connectivity solution establishes the association relationship between medical device data (parameters) and alerts (alarms) and stores and/or delivers the information as defined by the patient centric identification and association.

Today's Intelligent Hospital environment of care consists of many devices from a variety of vendors, resulting in combinations of devices, applications, and technologies to be integrated and utilized simultaneously. Achieving true integration and connectivity can be difficult, as data formats and encryption vary between vendors and device models [12, 13]. In order for medical devices to integrate with other medical devices and applications sharing data to support decision support and automated device control, such as alarm management systems, EMR, AI, and analytic software requires a robust, confirmed, or validated patient centric identification and association process to be in place. Once devices are associated, this aggregated association data can be archived and made available anywhere inside or outside the hospital. Advanced connectivity solutions enable the hospital to dynamically monitor all patients independent of their location ensuring redundancy and a standard level of safety across the facility.

Advanced Communications and Point of Care Data Delivery

Creating a seamless dataflow within the Intelligent Hospital is reliant on advanced communications methodologies and location-based technologies, as illustrated in Figure 1.14, to support and maintain rapid data collation,

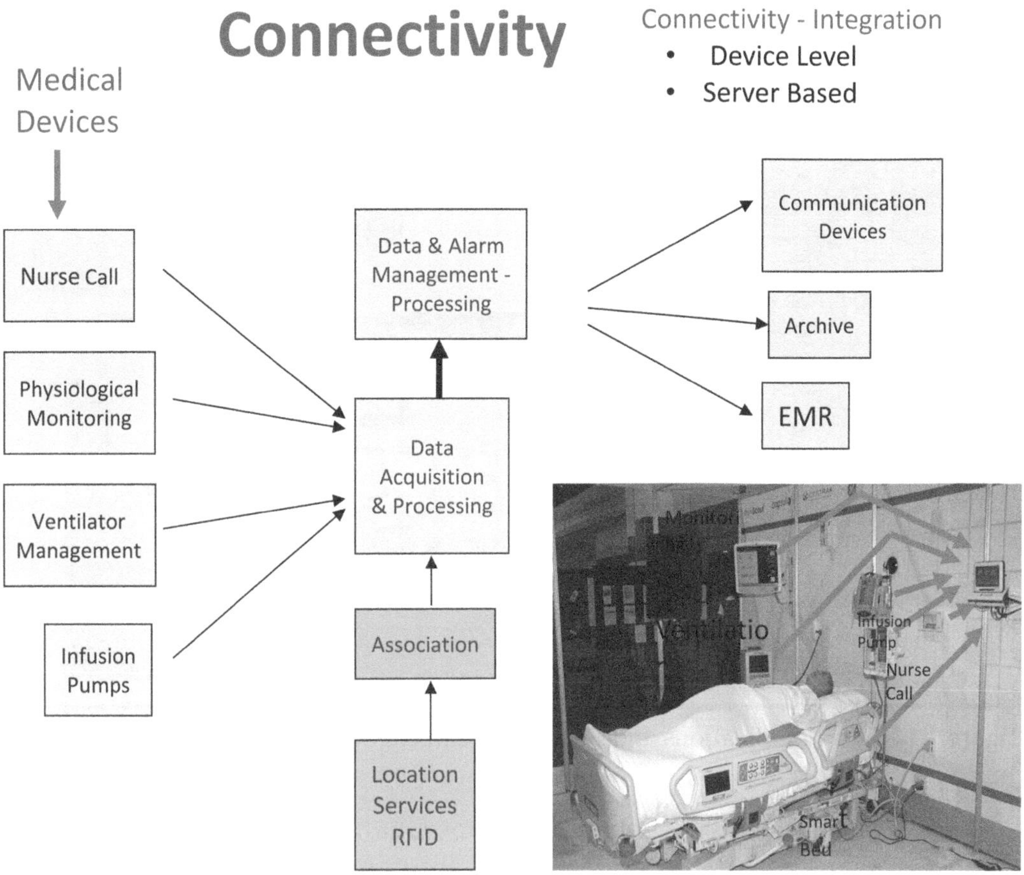

Figure 1.13 Connectivity.

processing, and dissemination enabling a dynamic capability for clinical consultation and decision support. As the acuity of the patient increases, both physician and nursing time at the bedside optimally increases proportionally. Therefore, the timely delivery of patient information, such as physiological data, alarms, laboratory results, or imaging, is key to optimizing patient treatment and care. With the deployment of a robust network infrastructure, connectivity, time synchronization, coupled with patient centric identification, the Intelligent Hospital is ready to implement and deploy point of care delivery solutions. The use of commercial handheld devices, such as iPads, wireless phones, and tablets, operating on the hospital's wireless infrastructure enables an optimal and cost-effective method for staff to receive and review information directly at the point of care while maximizing patient and staff bedside interaction. All that is required is that the handheld devices are associated with the staff member

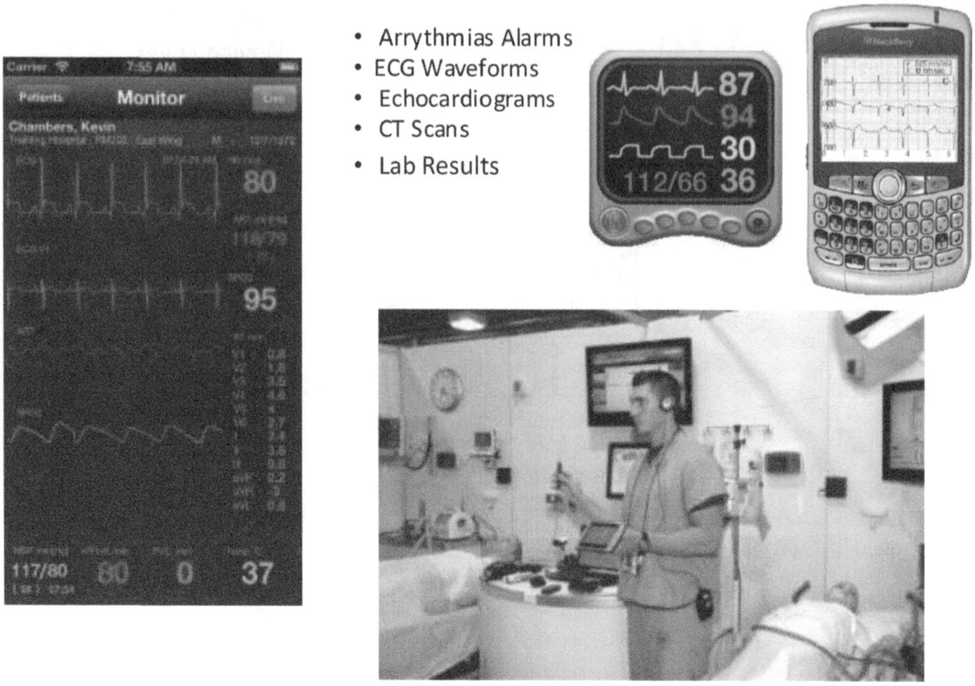

Figure 1.14 Communications and delivery of clinical data to the point of care.

who is associated with the patient and consequently the medical devices associated with the patient. All data flows through the hospital network via a connectivity solution enabling critical parameters and information to be rapidly routed from the medical device to the handheld devices. Additionally, using commercial communications devices which support voice over IP enables staff to staff communications, providing capabilities like bedside consultation, patient review, and treatment planning. Connecting these handheld devices through the institution's phone systems additionally enables incoming and outgoing phone calls to be routed directly to the appropriate party without being on hold or waiting for a call back maximizing bedside time and optimizing the delivery of patient information.

Location-Based Solutions

RFID Technologies

RFID is a wireless technology operating within the radio frequency spectrum (3 Hz–300 GHz) that accurately provides unique identification and location information of transmitting tags relative to the known position of

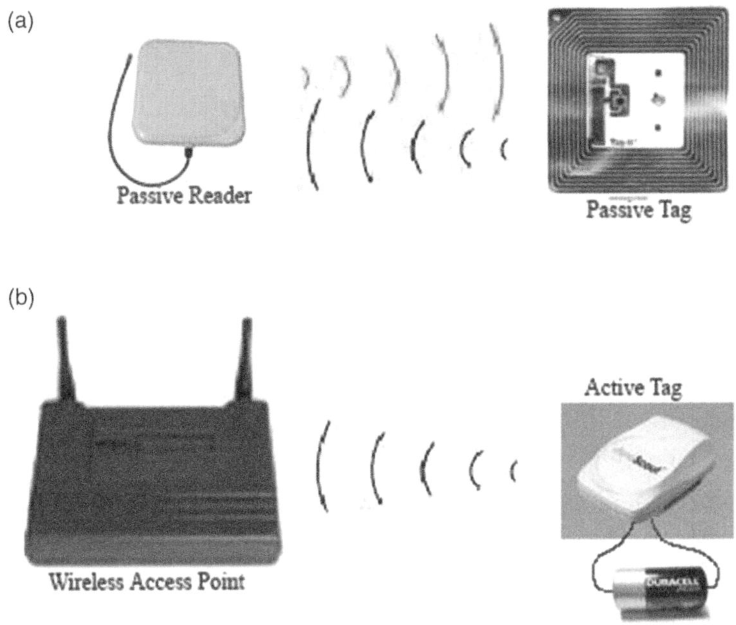

Figure 1.15 RFID technologies. (a) Passive RFID. (b) Active RFID.

multiple transmitting and receiving antennas. There are two primary types of RFID technology: active RFID, which contains and utilizes a battery to power and transmit signals autonomously; and passive RFID, which requires an external power source (antenna/transmitter) to provide power to the tag electronics and enable signal transmission, as illustrated in Figure 1.15.

Passive RFID

Passive RFID relies on a transmitting antenna to deliver energy to a receiving antenna built into the tag as illustrated. Received energy powers the passive tag which triggers a response signal uniquely identifying the tag's ID and associated asset. The advantages of passive tags include their small size, low price, the absence of internal power requirements, and their flexibility to be molded into many different shapes and forms. The main disadvantage is that passive tags have a limited or short read range which requires the installation of a dedicated infrastructure or use of short-range handheld transmitters/readers to function limiting their operational functionality as a global solution due to the high infrastructure costs. However, passive RFID is ideal for localized solutions, where antenna systems can be established to identify, monitor, and verify specific operations within specific locations.

Within the Intelligent Hospital, there are many applications for passive RFID technology including:

- Identification
  - Medical devices
  - Pharmaceuticals
  - Supplies
- Connectivity solutions
- Surgical tray assembly and tracking
- Surgical sponge and instrument tracking
- Pharmaceutical identification
- Sample collection
- Small transducer or device components inventory
- Patient centric identification
- Process validation
  - Sterilization, cleaning, or decontamination processes

Active RFID

Active tags are self-powered and have a greater potential for global or institutional applications and in many cases can utilize a facility's existing wireless infrastructure. However, they are larger, have a significant cost, and are dependent on battery power requiring periodic maintenance or battery replacement. An advantage of utilizing an active RFID solution is that specific use cases or applications can be implemented to support a variety of tag resolutions.

Levels of Granularity for Active RFID

The nature and requirements of the specific applications define the level of granularity and precision required in identifying the position of a tag. Tag visibility can be defined in terms of presence, location, and choke point architectures. Presence simply identifies that specific RFID tags are visible within the range of an access point or antenna; it is the least accurate of the location methods. This methodology is sufficient for general inventory or applications only requiring validation or verification (i.e., is the device present). Triangulation determines tag position based on a more complex mathematical method that utilizes multiple access points or receiving antennas. Using power measurements from the tag transmission and/or time of arrival statistics at multiple access points, the system calculates the location of the tag using triangulation algorithms. Typical range resolution is based on the geometry and density of the access points and the nature of the construction of the facility. Resolutions typically range from 10 to 30 ft. Choke-point applications utilize secondary hardware, such as low frequency transmitters, IR, or ultrasound transmitters mounted at known key locations such as doorways and exits. As a tag enters the "umbrella" or envelope as defined by the secondary hardware, a corresponding location ID is associated with the data string of the RFID tag transmission. This data is used by the RFID system in defining the precise position of the tag relative to the choke point hardware. Accuracy can be typically determined to within several feet based on the choke point distribution and configuration.

Leveraging of Existing Infrastructure

Active RFID technologies can require the use of a dedicated infrastructure consisting of a network of transmitters and receivers to provide location information, representing significant costs to an institution. However, most healthcare organizations already support a wireless infrastructure, such as 802.11, which can be leveraged to provide significant cost savings associated with an RFID deployment. Not all applications require room level accuracy, and consequently, the RFID can operate with a combination of mixed mode of resolutions which impacts installation costs. Many factors must be considered to determine the technology best suited for a particular healthcare environment, including the number of tags projected for deployment, whether the applications are localized or global.

Low Energy Bluetooth

An alternative next generation solution is Bluetooth Low Energy (BLE), a wireless, low-power personal or localized area network that operates in the 2.4 GHz ISM band. BLE connects devices over a relatively short range and was created for local IoT applications. A BLE device sends out a repetitive packet of information over one of three channels with random delays of up to 10 ms. BLE is used for applications that do not need to exchange large amounts of data and can also run on battery power for years with a reduced operating cost as compared to other location-based systems.

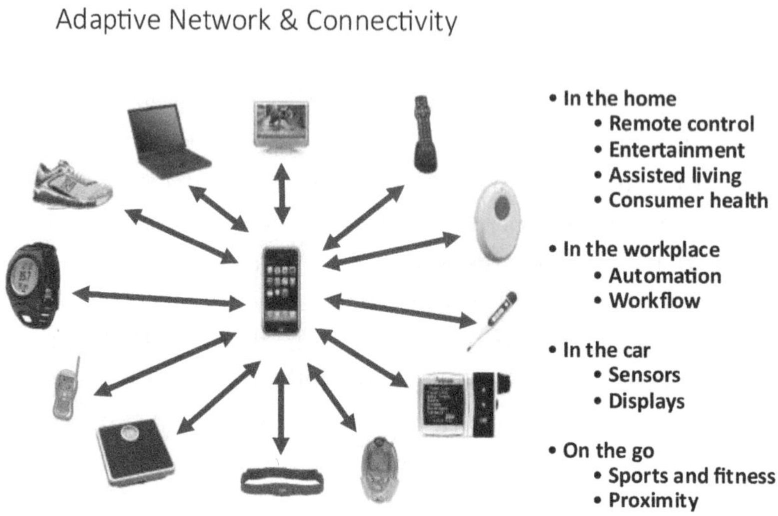

Figure 1.16 Bluetooth Low Energy (BLE).

BLE could be used for a broad set of healthcare-based applications including home automation, wireless medical devices, and exercise sensors. Some specific examples from medical devices are Smart Continuous Glucose Monitors, Bluetooth Enabled Pacemakers, Oximeters, Thermometers, Smart Treadmills, and Weight Machines. Receiving stations, antennas, or IoT devices such as cell phones establish a centralized client architecture providing a platform to receive BLE information analytics and visualization. BLE-based systems create an adaptive network enabling multiple BLE devices to dynamically share information with a central client as illustrated in Figure 1.16. An increasing number of medical and personal mobile devices supporting this technology have made it possible for hospitals to collect medical status and data to better engage their patients before, during, and after each visit.

Creating a Dynamic Institutional RTLS Solution

Within the Intelligent Hospital, an institutional location based RTLS solution design must be flexible and capable of supporting a combination of technologies including both active and passive RFID technologies, as well as next generation technologies such as BLE, while also supporting older legacy technologies, such as barcodes and IR (Figure 1.17). The overall solution will provide an archive of the locations of all tagged

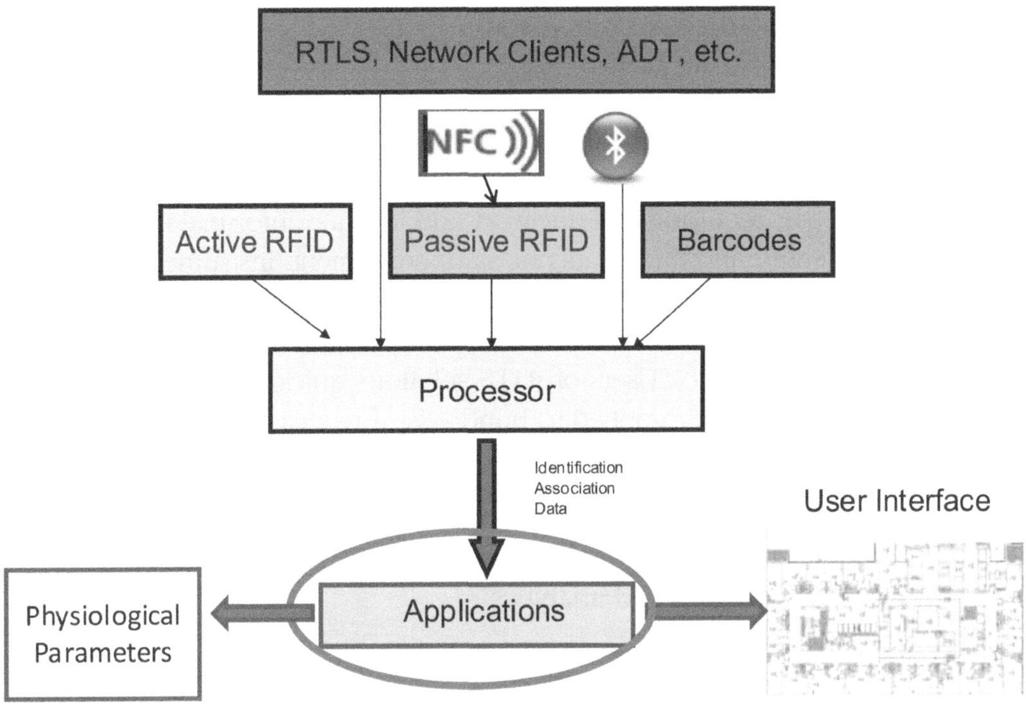

Figure 1.17 RTLS institutional solution.

assets, including patients, staff, medical equipment, IoT devices, supplies, pharmaceuticals, etc., enabling this information to be shared with clinical applications providing analytics which enable validation and the optimization of clinical and business processes and workflows, as illustrated in Figure 1.18.

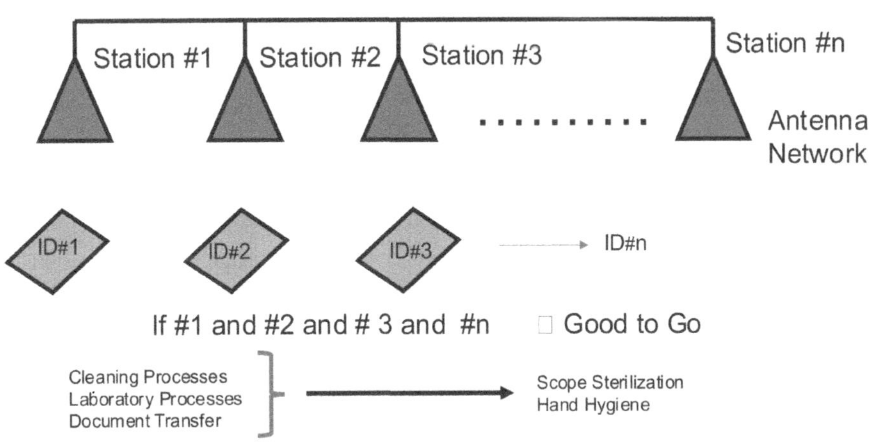

Figure 1.18 RFID technology solutions enable process validation.

RFID/RTLS Enabled Applications

RTLS solutions represent a significant enabling technology that establishes a unique identification and provides location information for a broad set of institutional assets. Assets can be broadly defined as almost any resource in a medical center requiring identification and location information. These technologies identify and track patients, staff, equipment, instruments, supplies, and pharmaceuticals. Coupling this information with existing processes and applications can help to characterize and optimize the workflow processes used in healthcare delivery. Users of RTLS solutions quickly realize that this technology can be used or applied to hundreds of use cases and applications serving a broad base of hospital operations as outlined below.

- Supply chain inventory management
- Medical device tracking
- Workflow monitoring and analytics
- Process validation
- Regulatory compliance

Supply Chain, Inventory Control and Management

Inventory management is perhaps one of the most obvious applications for identification and location-based technologies. Supply chain applications enable supplies to be uniquely identified, counted, assigned to specific locations, usage monitoring and scheduling for stock replacement, dynamically maintaining PAR levels, etc. Supply qualities can be closely monitored and verified to optimize cost reduction as well as ensure patient safety by ensuring supply availability and by checking and validating specific supply information, such as expiration, or existing recalls.

As illustrated in Figure 1.19, high-value supplies and implantable devices utilized in the ORs or other treatment areas such as the Cath Lab or Radiation Therapy are stored and managed via RFID coupled cabinets. The cabinet design incorporates RFID antennas to ensure all supplies in the cabinet are electronically visible. Each of these supply items is tagged with a unique RFID identifier (tag), and as the supply is removed from the cabinet for patient use or treatment, the inventory count is updated. The supply is also validated prior to patient use, ensuring it has not expired or has any FDA warnings or recalls associated. The device or supply can also be directly associated with the patient treatment used to enhance the patient documentation or associated with the patient for billing purposes.

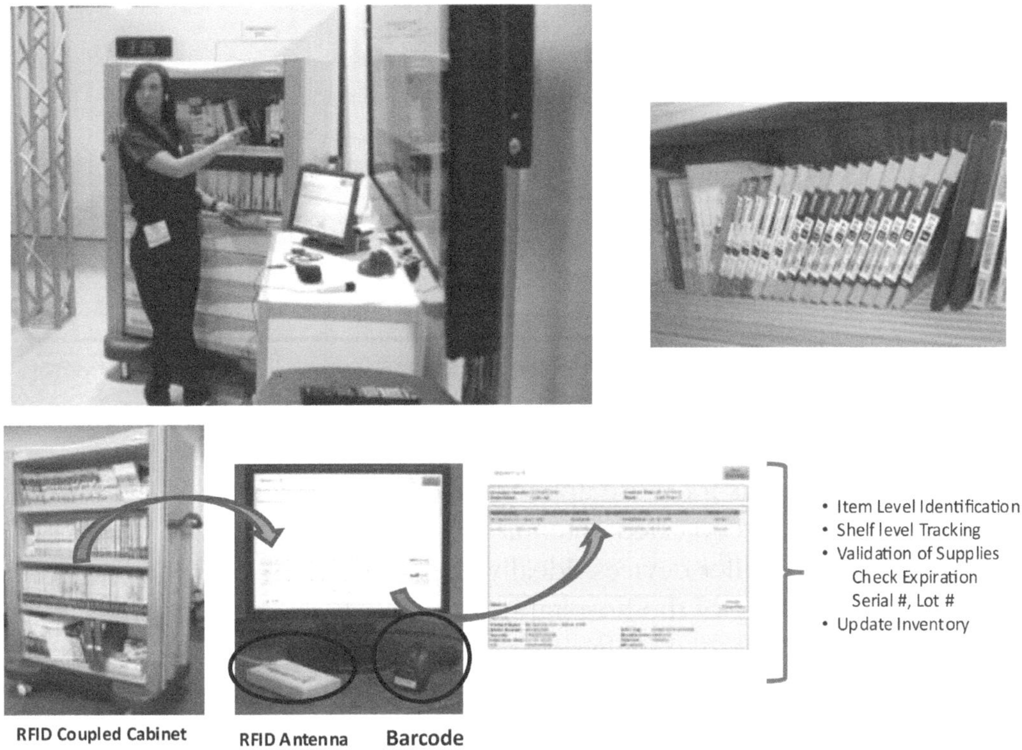

Figure 1.19 Supply chain RFID coupled cabinet.

Medical Equipment Inventory and Device Tracking

Management of medical systems, including devices, sensors, transducers, and support hardware, is also enhanced through the use of RFID/RTLS technologies. Regulatory agencies, such as the CMS and TJC, require Hospitals to ensure regulatory compliance, confirming device calibration and preventive maintenance at scheduled intervals to maximize patient safety and device integrity. The use of RTLS enables the Intelligent Hospital to uniquely identify medical systems and track and locate the system associated components to ensure all scheduled maintenance is performed in a timely manner as defined by the manufacturer requirements and schedules. Active tags are commonly associated with critical, highly mobile devices such as infusion systems, illustrated in Figure 1.20, providing global institutional visibility and real-time tracking.

However, many devices such as sensors, transducers, and probes are small and not conducive to large active tags. In this case, smaller form fitting

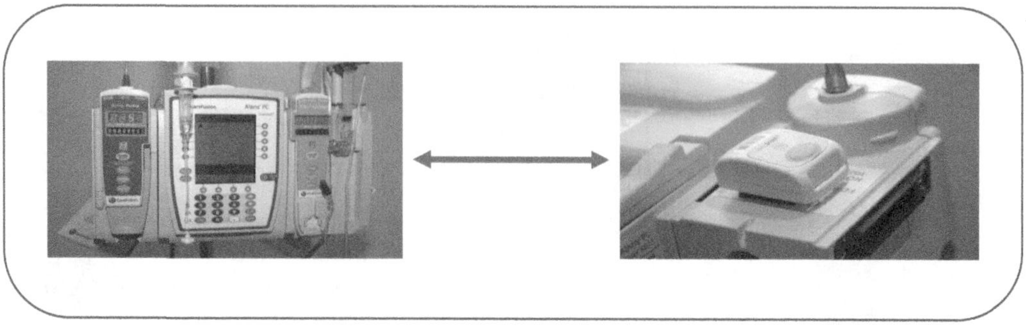

Figure 1.20 RFID coupled medical device tracking.

tags are advantageous for these applications, but as illustrated in Figure 1.21, handheld readers or dedicated antenna systems need to be used to identify and locate these smaller devices. Ideally, the information is merged to provide a global view of the hospital's critical assets or inventory. In this illustrated example case, the physiological monitoring system consists of many components, large and small, including the main computer, fixed base parameter modules, a variable configuration of specific parameter modules, a video monitor, mounting racks, etc. To uniquely identify all components of the system, a combination of active and passive RFID tags is used to ensure that all system components adhere to the regulatory requirements or guidelines. This use case specifically highlights an example of why an institutional solution needs to support multiple location-based technologies as part of the overall RTLS solution.

Asset Visualization

The use of RTLS enables medical devices and/or other assets to be dynamically tracked and provides a global view of the institution's assets within the patient care areas down to the room and bed level. The ability to further associate device location with device status significantly enhances the quality and value of the search processes. We can now specifically identify and target devices ready for use and bypass those currently in use or unavailable.

Enabling status information to be coupled to location within the RTLS system can be accomplished in multiple ways. The optimum design approach focuses on developing a direct integration with a medical device to extract various operating parameters, such as operating status, clinical parameters, or alarm information. This is a complex development effort and

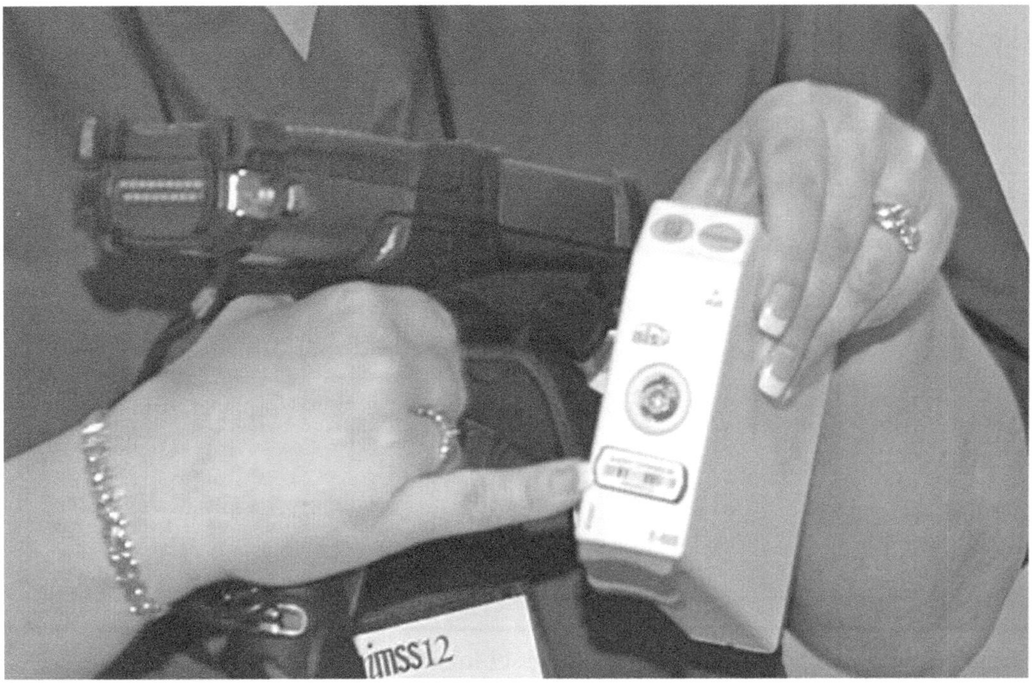

Figure 1.21 Passive RFID identification.

is difficult to perform across a broad base of medical devices or systems. However, for specific medical devices which are commonly associated with frequent searches and are in high clinical demand, such as infusion systems, this provides significant added value. Based on a collaboration between Memorial Sloan Kettering Cancer Center, Stanley RTLS, and BD Carefusion, such an integration was developed and deployed. Device status (infusing) was extracted from the pump system and associated with the device icon in the RTLS mapping or display.

An alternate approach is to utilize an event button commonly available on many active RFID tags. The event buttons can be used to identify the specific state of a device and can be redefined to represent a variety of functionality, such as in use or available, needs service, etc.. This method does present some limitations since it does require supporting workflows to ensure the state of the device or status is accurately set. The accuracy of the information becomes highly dependent on the compliance of the staff to set or use the event button correctly. However, this solution is a significantly simpler solution and is readily available and can be applied to a broad base of devices or other assets without additional cost or any significant development effort. In terms of visualization, devices can be color coded on

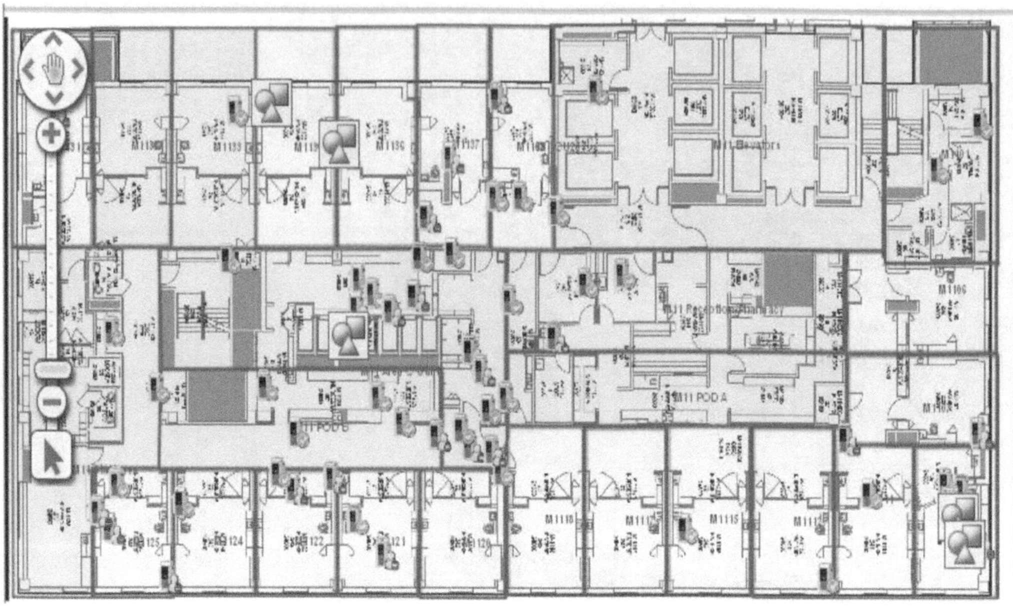

Figure 1.22 Asset tracking: identifying asset location and operational status.

the floor map display defining equipment locations with status, for example, defined as red being unavailable or green as available. This approach provides a robust and accurate representation and view of device status as illustrated in Figure 1.22.

Device status coupled with color coding can be defined to support many device conditions or statuses, optimizing workflow or device management:

- In-use/Available
- Clean/Dirty
- Needs repair/Operational
- Incident occurred
- Requires preventative maintenance

Utilization Statistics

Once device status and availability is defined as a function of time for each medical device, it is possible to compute the overall utilization or uptime/downtime of medical devices. This information is extremely valuable and can be used by the administration to make strategic planning or purchasing decisions, establish or update device allocation and distribution decisions, and optimize the overall medical device supply chain (Figure 1.23).

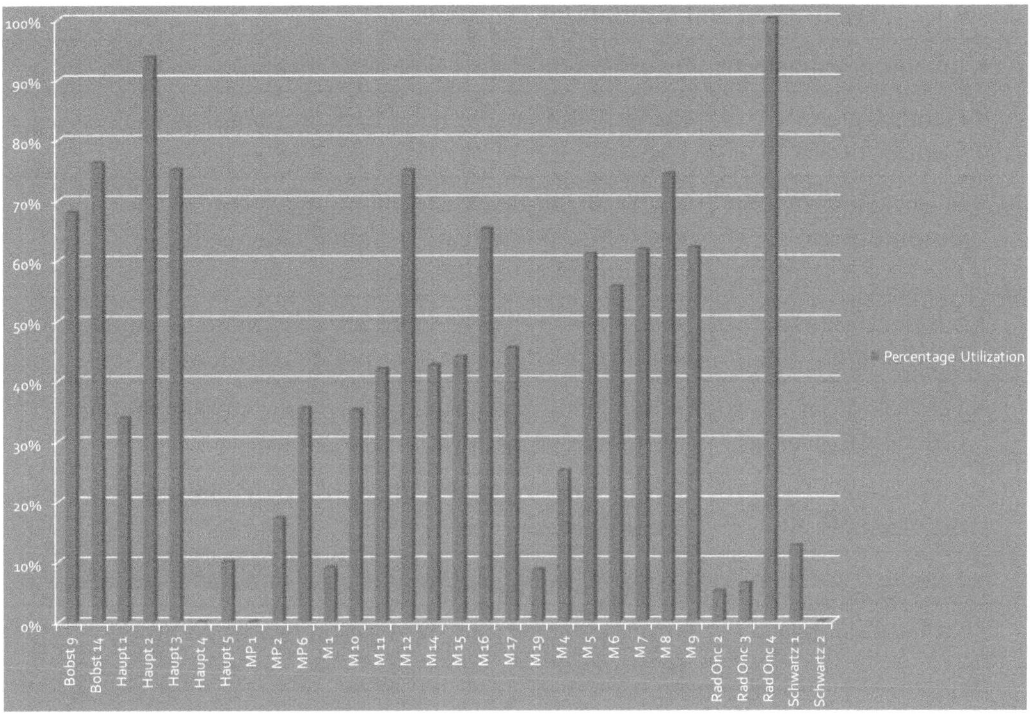

Figure 1.23 Device utilization by care unit.

Operationally, this information can be trended over time, to represent the dynamic changes in utilization and establish an optimized PAR level, establishing specific or required device volumes as a function of day, operational time of day, etc.

This process is applicable to most devices; however, the devices typically associated with this type of monitoring focus on highly mobile or critical care devices as outlined.

- Infusion systems
- Portable C-arms
- Wheelchairs and stretchers
- Critical care devices
  - Code carts/defibrillators
  - Ventilators
  - Anesthesia machines
- Beds
- Stretchers/Wheelchairs

Table 1.1 Verification and Validations Processes

• Inventory Validation	Ensure Medical device or asset inventory accuracy
• Hand Hygiene Compliance	Monitor and ensure hand hygiene compliance
• Surgical Instrument Confirmation	Ensure all Surgical Instruments are accounted for and recovered minimizing patient incidents due to instrument retention
• Surgical Sponge Verification	Ensure all surgical sponges are accounted for minimizing surgical incidents due to sponge retention
• Verification of Cleaning Processes	Ensure all devices undergo a specific sterilization or cleaning process prior to patient use
• Medication Administration	Confirm the 5 patient rights prior to medication administration
• Sample Collection	Ensure all samples are associated with the correct Patient
• Process Validation	Ensure cleaning and sterilization processes are performed

Process Verification and Validation

One of the critical and valuable operational capabilities of location-based solutions is enabling verification, validation, and quantification of critical processes and assets (Figure 1.18). Location technologies such as RFID/RTLS and BLE provide a mechanism to validate or confirm that specific processes have been performed. We have already highlighted several examples including supply chain management and hand hygiene compliance; however, there is a broad base of applications that can be implemented to minimize errors, ensure patient safety, and demonstrate regulatory compliance as outlined in Table 1.1.

Requirement for Inventory Accuracy

All healthcare institutions struggle to maintain an accurate inventory of their assets. Devices can enter and leave the institution via various pathways. In the case of medical systems and devices, there are regulatory requirements to ensure an accurate inventory, used to schedule and manage preventive maintenance, calibration, and updates ensuring patient safety. As part of a hospital's accreditation process, TJC specifically reviews how an institution maintains the critical inventories including medical devices, safety devices, and specific supplies. A hospital's medical device inventory is a dynamically changing entity, constantly dealing with the onboarding and retirement of devices, along with the management of a variety of temporary devices including incoming loaner or rental devices, 3rd party owned devices, patient owned devices, and telemedicine home based devices. The use of active and passive RFID enables inventory to be dynamically validated. This can be accomplished using an antenna infrastructure coupled with the use of handheld readers where at predetermined intervals a hospital can be scanned to confirm or validate the inventory. This process validates device presence and specifically focuses on determining the status of inventory that is no longer visible. The accuracy of the validation process is dependent on an institution's onboarding guidelines and practices ensuring all devices are labeled, tagged, and inventoried in accordance with a hospital's guidelines and standards.

Workflow Metrics and Optimization

The ability to quantify workflow processes and define the corresponding impact on patient care and safety provides valuable data and establishes a pathway for continuous process improvement. Real-time tracking of patients and staff and correlating their location with specific events provides many options for workflow monitoring, management, reporting, and optimization. One specific metric of interest is the response times of staff to clinical events, including patient calls, codes, critical alarms, etc. Availability of this type of data provides quantitative data and metrics on relationships, such as the extent or level of patient-to-staff interaction. This data provides a direct correlation to quality patient care and an indicator of potential patient satisfaction. This information can also be correlated to staffing requirements used to define and maintain an institution's staff ratios and optimal level of patient care and interaction. At times, patients are not fully aware of the specifics of their care and surroundings or even the sequence of events, due to treatment, drugs, or their clinical condition. Occasionally, patients convey misinformation to family members about not having been seen by the nurses or physician for prolonged periods of time, leading to questions about the quality and level of patient care being provided. Using RFID/RTLS workflow solutions provides specific metrics to monitor staff response and can be correlated to specific events including patient interaction.

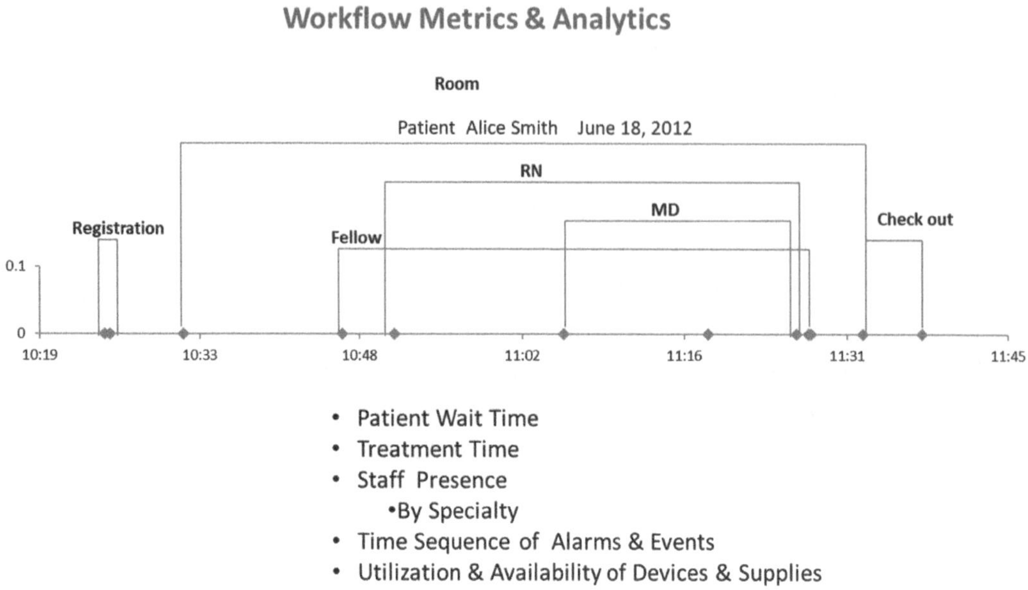

Figure 1.24 The patient experience.

This provides hard data as to the number and duration of nursing or staff interactions within the patient room or bedside. This same data can be used by administration and regulatory staff to confirm or validate that the quality or standard of care is being provided – all very useful information.

As illustrated in Figure 1.24, based on a study performed at Memorial Sloan Kettering, specific areas of the hospital were outfitted with localization hardware to enable a choke point architecture defining specific locations such as registration, waiting area, treatment rooms, etc. The ability to quantify workflow processes through these instrumented locations generated detailed information on the timeline of clinical staff and patient interaction. The data clearly identifies patterns in the patient experience and provides metrics on potential improvements and optimizations. These types of real-time metrics provide valuable and dynamic data and establish a pathway for process review, enhancement, and improvement.

Global Visualization of the Intelligent Hospital

Looking at this data from a broader perspective, synchronizing the patient and staff location information with real-time physiological events and clinical activities paints a clearer picture of real-time patient care dynamics.

Figures 1.25 and 1.26 illustrate various ways in which this new enabling dynamic information can be used to identify the overall interactions and staff utilization through the entire patient care area or unit. The ability to provide this dynamic global view of the operations within the Intelligent Hospital enables many processes and resources to be monitored simultaneously as outlined below:

- Room/Bed availability
- Staff location and availability
- Patient-to-Staff interaction time
- Emergency situations (alarms, patient falls, etc.)
- Medical equipment location, status, and utilization
- Discharge and cleaning

Visibility in the Intelligent Hospital

As has been highlighted throughout, enabling technologies have been coupled with a variety of data adding a new dimension in understanding

Figure 1.25 Workflow management and process improvement.

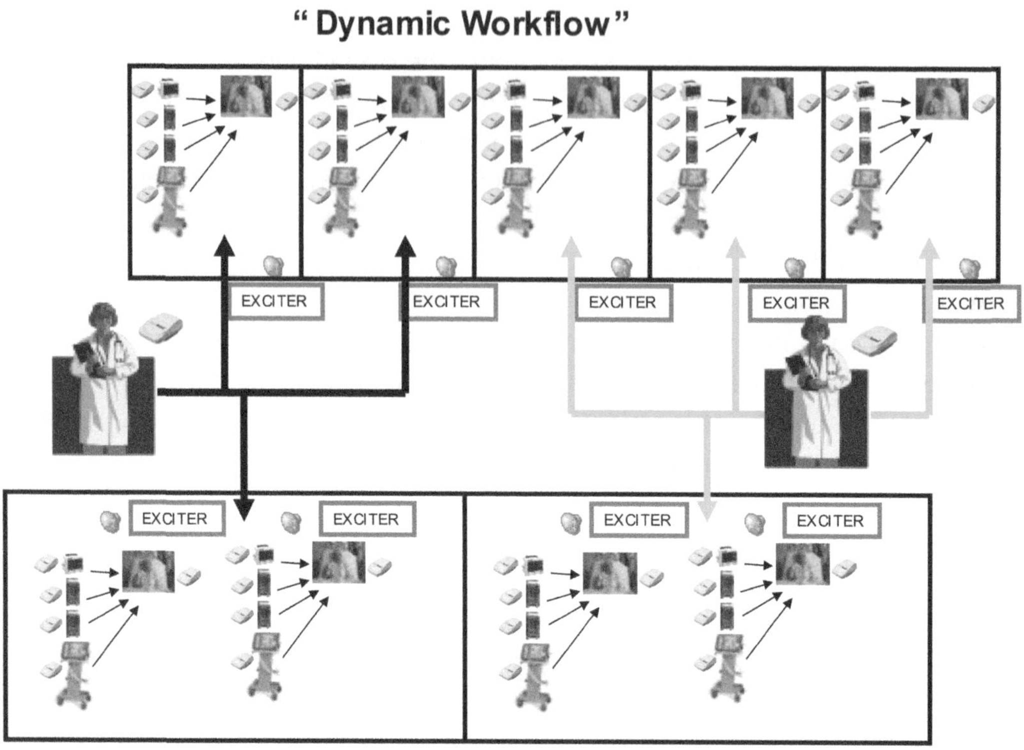

Figure 1.26 Synchronized RFID tracking percent (%) event streaming.

and defining the overall patient care process. This dynamic visibility is an important feature or component enabling constant assessment and optimization of the process across the overall Intelligent Hospital. These correlated datasets provide the ability to aggregate information from varying sources creating new and dynamic information to monitor and visualize a patient's status. Data presentation can be tailored or customized as a function of location and by process. As an example, the data of primary interest in the ICU may focus on physiological parameters, and in telemetry areas patients' location may be of increased importance. Information can be centralized in specific war room or control room settings, where a complete overview, as well as a detailed review and monitoring, is possible from a global perspective. This type of environment has traditionally been used for dedicated and focused monitoring areas such as telemetric physiological monitoring units but can be expanded to monitor and provide metrics on many patient care delivery processes.

Conclusion

In this chapter, we have outlined examples of how enabling technologies can directly enhance patient care and optimize the business of how care is delivered. These technologies, coupled with advanced medical systems and clinical applications, provide a new and combined view of patient care and overall care delivery. These enhanced processes enable checks and balances, and verification of the care process, and directly impact the reduction of errors maximizing patient safety and overall satisfaction. In this section, we have touched on several technologies and their potential impact on the care processes; however, there are many additional technologies which have a positive impact on the care processes including clinical software to provide decision support and prediction, such as AI. New advanced medical systems are being designed to directly incorporate features to add new visibility and perspective to the care process, including imaging enhancements coupled with robotic systems, such as surgical systems, and targeting systems including lasers. The Intelligent Hospital concept does represent a continuously expanding and changing environment with need for medical systems coupled with an ever-expanding set of enabling technologies. Future editions of the book will continue to highlight the evolution of enabled patient care processes and their impact on improving patient outcomes and patient safety and enhancing the overall patient experience.

References

1. DeGroot, H., Patient classification system evaluation, Part I: essential system elements. The Journal of Nursing Administration 1989; 19(6):30–35.
2. Van Slyck, A., Johnson, K.R., Using patient acuity data to manage patient care outcomes and patient care costs. Outcomes Manage 2001; 5(1):36–40.
3. Arvantes, J., Emergency room visits climb amid primary care shortages, study results show, AAFP News, 27 August 2008, http://www.aafp.org/online/en/home/publications/news/news=now/health-of-the-public/2008
4. Frisch, P.H., Pappas, H.P., Leveraging Technology as a Response to the COVID Pandemic, Routledge, 2023, ISBN 97803677693.
5. Morgan, S., Medication error statistics. The Prescription 2005; 1(1), July. Available at http://www.msnbc.com/id/7995137/
6. Phillips, D. P., Christenfeld, N., Glynn, L. M. Increase in US medication-error deaths between 1983 and 1993. Lancet 1998; 351:643–644.
7. Cardinal Health: Statistics – Medication Safety & Education.

8. Joint Commission Patient Safety Goals, http://www.jointcommission.org/PatientSafety/NationalPatientSafetyGoals/09_hap_npsgs.htm
9. Frisch, P.H., Pappas, H.P., Leveraging Technology as a Response to the COVID Pandemic, Routledge, 2023, ISBN 97803677693, Chapter #2: Optimizing Infection Control & Hand Hygiene during COVID.
10. Dyro, J., Clinical Engineering Handbook, Academic Press, 2004.
11. David, Y., Maltzahn, W., Neuman, M., Bronzino, J., Clinical Engineering, 2003: 432. https://doi.org/10.1201/9780203009178
12. Frisch, P.H., Miodownik, S., Booth, P., Carragee, P., Dowling, M., Patient centric identification and association. IEEE Engineering in Medicine and Biology Society, Proceedings of the 31th International Conference. 2009: 1722–1725.

13 Day, B., Standards for medical device interoperability and integration. Patient Safety & Quality Healthcare, Jan/Feb 2011.

Section 2

Technology-Enabled Workflows across the Environment of Care

Introduction

The Intelligent Hospital consists of a variety of patient care environments to provide triage and assess a patient's status, perform diagnostics, provide treatment, and manage the overall patient care. In addition, the hospital includes a broad base of support functions, such as biomedical engineering, information technologies, facilities management, food services, and housekeeping. Hospital operations can be organized in a variety of ways, which vary from institution to institution but can be generally outlined as illustrated in Figure 2.1, dividing operational functions into informational, therapeutic, diagnostic, and support services.

These functions are performed and distributed over a broad spectrum of patient care areas and supported by many distinct clinical operations and services.

As highlighted, the overall objective of the Intelligent Hospital is to establish a seamless continuum of care that enhances patient outcomes and care and optimizes patient safety, patient satisfaction, and the overall experience of the patient as they transition through the healthcare system. The patient experience is directly related to the specific processes associated with each of the care environments, which can occur in phases of the patient care process occurring at different locations experienced over many visits

DOI: 10.4324/9781003414568-2

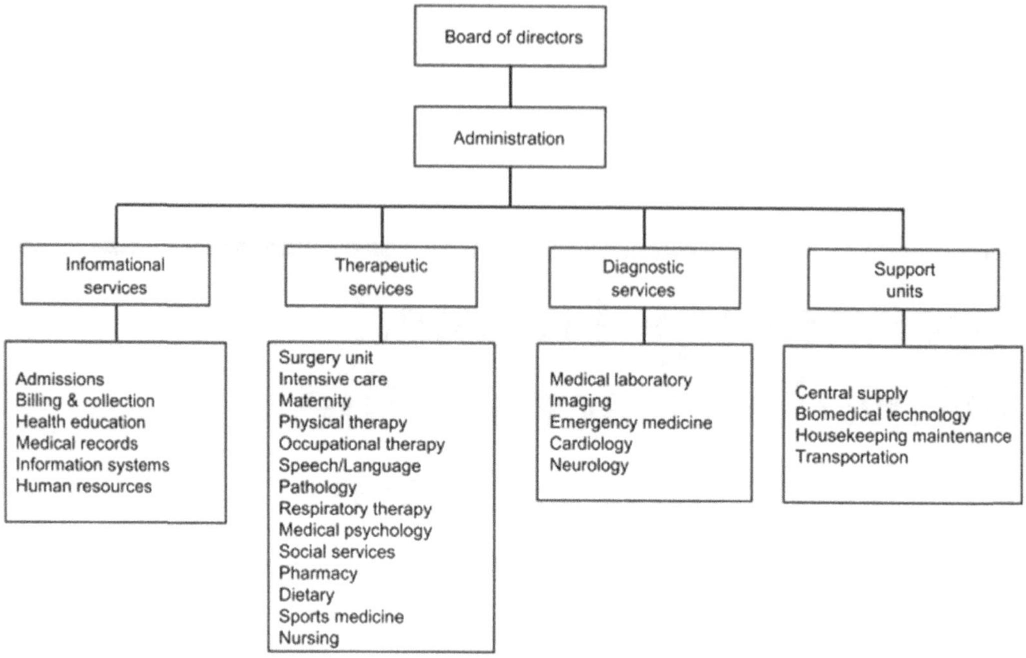

Figure 2.1 Organizational structure of operations within the Intelligent Hospital.

or sessions. As an example, the patient experience can include emergency scenarios, such as ambulance and/or Emergency Department (ED) visit; diagnostics, such as laboratory tests, and imaging, which can include a variety of imaging modalities including CT, MR, PET, and ultrasound; treatments which may be invasive including surgery or cardiac catheterization; treatments that occur over several cycles, such as Radiation or Chemotherapy; and patient care experienced in the critical care areas (ICU, Step Down) and in general patient care areas. Follow-ups can occur in the hospital or more commonly occur in regional ambulatory care setting locations with an increasing number of cases extending to include medical office visits, as well as remote telemedicine applications used in the home (iHome). How the patient and their family members perceive their care process is based on many factors, including outcome, effectiveness, satisfaction, and confidence in the quality of care associated with their overall care process.

The patient's care process consists of many individual workflows, occurring within many locations and facilities within the health system. The goal of the Intelligent Hospital is to establish a care model where the standard and quality of care process is consistent for all aspects of care across the entire continuum of care. The patient care model consists of many clinical workflows associated with the many processes that make up the patient care process. The critical goal is to optimize patient care, while ensuring that all relevant patient information associated with these processes is seamlessly acquired and available at all stages of the patient's treatment and care scenario.

Highlighting how diverse enabling technologies, integrated and used to create optimized patient care workflows, is critical to understanding the Intelligent Hospital concept and its impact on the patient care process within various care environments. These advanced workflows optimize safety, validate information integrity and accuracy, and provide an enhanced patient management process and experience.

Technology-Enabled Workflow

Workflows vary from institution to institution based on the internal best practices, the institution's policies and procedures and the types of medical systems, clinical applications, enabling technologies, and infrastructure deployed within the healthcare system. The patient care workflows consist of and exploit the many specific use cases made possible using enabling technologies. All healthcare institutions use a common set or type of medical devices (physiological monitoring, infusion systems, ventilators) coupled with institutional clinical applications to support electronic medical record (EMR), medication administration, sample collection, patient centric identification and association, analytics, and decision support. It is important to note that within the Intelligent Hospital's patient care environments, medical devices are no longer stand-alone devices operating asynchronously but are now interfaced with other medical devices and integrate with a variety of clinical applications, IoT devices, and enabling technologies to create complex systems. These complex systems provide real-time metrics and parameters of the patient, advanced analytics, decision support, data archival, visualization, and data delivery. This overall functionality creates opportunities to provide enhanced patient care safety with processes that also optimize clinical and business operations. Unique patient care environments such as the OR, Clinical Laboratories, Pharmacy, Radiology, and Radiation Oncology support unique use cases and workflows. As an example, in the OR, tracking surgical sponges and RFID-coupled surgical instruments enable identification and monitoring to ensure patient safety and confirmation of sponge and instrument recovery. In the Labor and

Delivery Unit (LDR), monitoring devices and applications are focused on infant monitoring to ensure the baby is not under stress. Additionally, supporting areas like Biomedical Engineering and Central Supply utilize specific technologies and applications to confirm and validate cleaning and sterilization processes and workflows ensuring medical device infection control, operational integrity, and patient safety.

As highlighted throughout Chapter 1, to create a standardized set of processes and workflows that share or provide information across the entire continuum of care of the Intelligent Hospital, several institutional technologies and the following processes need to be standardized throughout the Intelligent Hospital to ensure consistent data availability and integrity:

- Network and wireless infrastructure
- Time aynchronization
- Standardized and unique medical device naming conventions
- Unique naming convention of facility locations
- Patient centric identification and association
- Data archival

In summary critical to all workflows, the Intelligent Hospital combines a robust network and wireless infrastructure, coupled with real-time operational intelligence from multiple location systems, RTLS/RFID/BLE systems, with integrations to a common set of clinical applications and archival software systems all coupled with several modalities of communication and visualization. This enables and provides a rapid enterprise-level interaction available on a variety of dashboards, on a portable tablet or a handheld device. The Intelligent Hospital also deploys a network infrastructure supporting a Wi-Fi-based time synchronization system using Network Time Protocol. This guarantees that all clocks, NTP-enabled systems, and medical devices are synchronized throughout the entire hospital and commonly documented throughout all systems.

Advanced Workflows

The following examples highlight several advanced workflows and define how each can be created, validated, and operationalized by exploiting multiple use cases enabled through the integration of diverse technologies, medical devices, clinical applications, analytics, and common infrastructure to create a standardized, safe, and optimized environment care experienced during all phases of a patient's treatment and care. The following workflows are fictional; however, they are representative of realistic and common patient care scenarios where specific use cases and workflows are applied within specific patient care environments. The following workflows are focused to demonstrate how enabling technologies, clinical applications, and integrations can directly enhance patient care and outcomes, increase patient safety and satisfaction, while also optimizing clinical and business processes in a variety of patient care environments and scenarios.

- Emergency response and triage
- Ambulance/Emergency department
- Surgical suites/OR workflow
- Cath lab – Cardiology workflow
- Critical care – ICU workflow
- Patient care management workflow
- Pharmacy workflow
- War room workflow

Emergency Response and Patient Triage Workflows

Ambulance Transport and Triage Workflows

Very frequently, the patient process starts in an ambulance responding to an emergency call due to an accident or unexpected medical emergency or event (Figure 2.2). Critical to patient outcome is providing a rapid and optimized response to the injured or ill patient. The ability to quickly assess their condition and provide immediate medical treatment maximizes the probability of a positive outcome. Receiving patient physiologic data, demographics, and real-time status while the patient is in transport can significantly impact the response time and treatment delivery, increasing the likelihood of a positive outcome. Technologies enabling cellular communications and data transmission, such as Remote Telemetry from Ambulance, enable 12-lead ECG data, along with vital signs including invasive pressure, temperature, capnography, SpO_2, and cardiac output acquired through the ambulance defibrillator to be directly transmitted to the Emergency Room. These systems can provide a variety of analysis

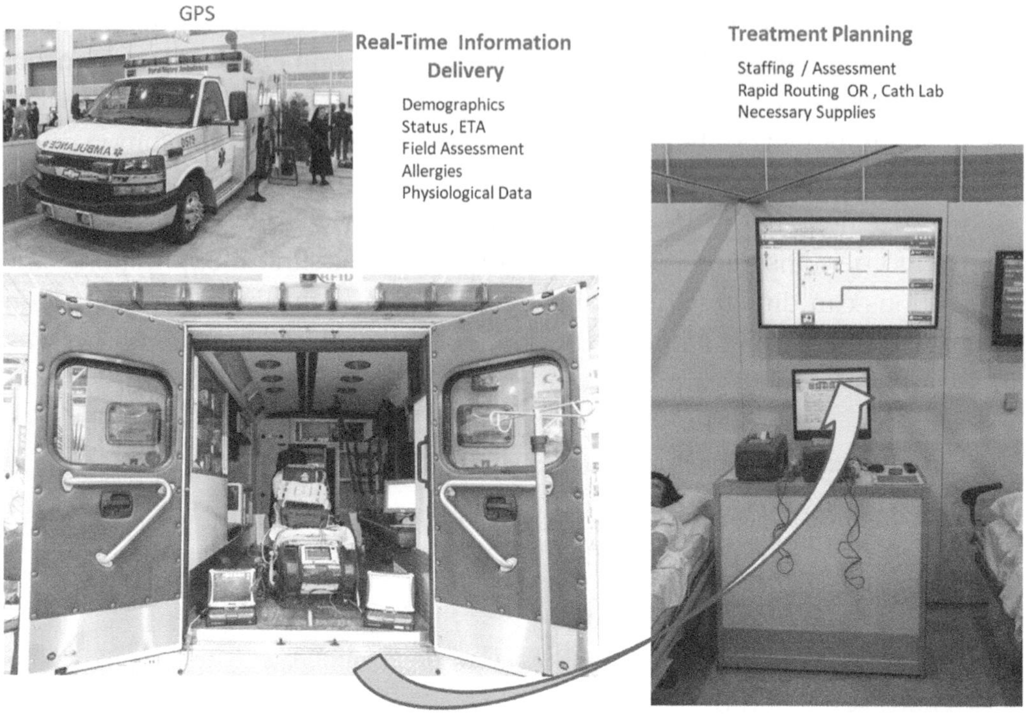

Figure 2.2 Ambulance emergency response and triage.

algorithms to interpret ECG waveforms and provide enhanced information on arrhythmias, such as Afib, Vtach, or ST segment elevation. This information can be directly displayed and available in the ED along with the ambulance route and estimated time of arrival. By networking the EMS 12 lead ECG data and supporting analytics to the hospital network, it enables the system to populate hospital-based information systems such as the EMR and cardiology management system for remote viewing and future comparison. In the case of an existing patient (patient with an existing hospital medical record), these systems enable direct comparison to previous patient histories and comparison to previous physiological information. With a direct combination of voice and data with the Ambulance, consultation allows the ED to prepare in real time for a patient arrival or make critical decisions, such as directly routing the patient to other treatment areas, such as the Cath Lab or OR. Critical to providing a rapid response is receiving patient physiologic data and demographics while the patient is in transport which can significantly impact the response time and treatment delivery, increasing the likelihood of a positive outcome.

Emergency Department Workflows

In a typical Emergency Department (ED) scenario, multiple patients arrive or are en route via ambulance, resulting in a steady stream of patients arriving at the ED having a broad range of issues and acuity. The ED is a high-pace environment requiring rapid triage, diagnostics, assessment, and transportation of patients to appropriate treatment or monitoring areas of the hospital (Figure 2.3). A critical and very common issue within the ED is patient overflow. To manage large numbers of patients and avoid diversion to other hospitals, location-based technologies, such as RFID/RTLS, are used to monitor patient throughput and to establish and define virtual locations within the ED. In scenarios as recently experienced during the COVID pandemic, patients are often doubled up and tightly placed in any available location, within the corridors and even overlapping within the patient bays. To effectively manage patients, these virtual locations must depend on patient centric identification and association establishing

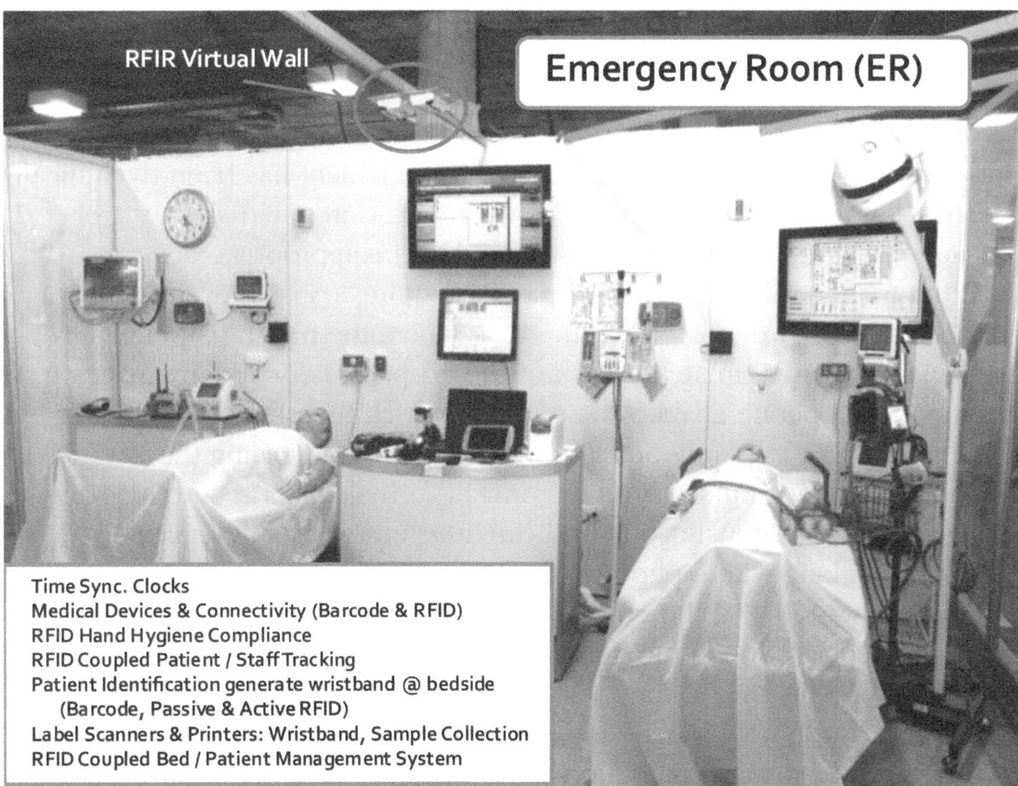

Figure 2.3 Emergency department.

dynamic relationships between the patient, staff, equipment, supplies, and pharmaceuticals. These applications require high accuracy RFID locating typically down to a very specific footprint. Dedicated Location Transmitters are used to receive signals from the RTLS tags enabling high resolution location definition, separation, and tracking of equipment and other assets enabling the ED to flex its capacity at critical times of overflow. The technology deployed to provide the identification, association can be based on a variety of location technologies including active or passive RFID, IR, and low energy Bluetooth.

Environmental Monitoring

In the event of a pandemic such as the Coronavirus, isolation and environmental monitoring, temperature, humidity, and particulates become increasingly important. Clinical applications enable dynamic customized assessments to confirm exposure or presence of specific viruses or diseases and dynamically triggering notifications to efficiently sort patients and alert the appropriate agencies. The assessment process utilizes learning algorithms to confirm if the patient has fever, sore throat, fatigue, has recently traveled abroad, etc. These learning algorithms match answers to specific known patterns, and if the patient fits the patterns, an alert to quarantine can be automatically identified and/or triggered. If the patient has been dynamically identified as a carrier of a contagious virus (i.e., Coronavirus), they are isolated, and the information is distributed to the appropriate agencies.

If the triage process confirms a patient requires consultant from another service such as Cardiology, as in the case where the patient is potentially having a myocardial infarction, use of the hospitals networked wireless devices or cell phone enables the nurse to search the physician schedule for a cardiologist currently on call for the ED without leaving the patient bedside. The triage nurse sends the on-call cardiologist a secure message explaining the current situation and can attach the patient's clinical information along with a photo of the ECG. She can also select to have the system notify her if the on-call physician does not review the message within a predetermined time frame. The on-call cardiologist receives a notification on his smartphone that he has an urgent message waiting in the system. Upon opening the message, the physician quickly reviews the nurse's secure message, the patient's clinical information and the ECG. The physician responds to the message, immediately directing the nurse to alert the cardiac catheterization lab staff of the incoming patient and administer necessary medications. The time from nurse initiation of contact to physician response is less than 5 minutes. This rapid assessment time can have a critical positive impact on the patient's outcome.

Based on the policies and protocols of the institution, the patient can also bypass the ED completely and be directly routed to the Cardiology Suite based on the assessment process obtained from the ambulance. Within the Cath Lab, this enables the necessary clinical staff to be ready before the patient optimizing patient treatment. Having this optimized workflow and procedure in place has enabled the hospital to reduce their door-to-balloon time to less than 90 minutes as recommended by the Joint Commission, which significantly impacts the potential for a positive outcome.

To ensure the readiness of the ED staff to manage incoming patients, an electronic dashboard provides a global visualization of the ED and enables the nurses/physicians to plan for incoming patients, monitor existing ED patients, and review physiological data and images of ECG waveforms for the patients. The ED staff can also confirm the availability of all required supplies and optimize the ED supply chain management processes. General supplies can be managed using the two-bin RFID system, which ensures optimum stock levels and demand accuracy. This practice eliminates the need for manual demand assessment rounds by automatically triggering replenishment requests when a predefined batch of an RFID-coupled product has been depleted. The nurse can also select a patient on the touch-enabled kiosk or use a barcode or RFID reader to identify the patient and associate any RFID-enabled supplies for charge back to the patient.

Other arriving patients may have multiple orthopedic injuries. As before trauma surgeon is notified to arrive directly at the ambulance bay to assess the patient and determine whether surgery is needed. If surgery is required, the surgical schedule and surgery preparation start can trigger in the ED and send the patient directly to the OR for select surgical procedures. The triage or OR prep nurse can additionally confirm that two small fracture kits are available in the OR using the RFID-coupled supply chain management system.

Patients and Staff Identification

As patients arrive at the ED, they are placed in bays pending assessment, pain management, and rapid triage. In this high-pace environment, infection control is critical requiring hand hygiene between the entrance of the patient bays. An RTLS badge uniquely identifies staff via the RTLS reader correlating whether that tag's presence was sensed at the hand-washing dispenser. If the care provider enters the patient area without following the hand hygiene protocol, the RTLS badge provides a signal/reminder upon entry, allowing the clinician an opportunity to self-correct. Returning to the hand sanitizer dispenser and using the dispenser, the embedded dispenser module positively identifies the clinician and sends a wireless signal to the badge which cancels the alert. Hand Hygiene Compliance Monitoring software applications provide administration and leadership with real-time validation and metrics, allowing hand hygiene compliance data to be evaluated over time.

Automate Patient Registration

To expedite a quick patient registration at the bedside, a variety of techniques can be used. If the patient is not in the hospital system, a unique identifier can be defined utilizing the driver's license, coupled with a bedside label printer, which enables a temporary wristband to be printed directly at the point of care. The patient wristband has a combination of barcode, passive RFID, and text information. An active RFID tag can also be added enabling visualization and tracking, both locally and institutionally. This unique temporary ID process can be reconciled to the patient's EMR at a later time.

As the nurse enters the ED, her RFID-based presence integrates with the nurse call system automatically showing that a nurse is present in the room. Within the ED, the close proximity of beds requires bay-level location accuracy to ensure the unique ID of the patient, devices, supplies, and pharmaceuticals are associated with the right patient. To do this, a virtual wall or divide can be created using RTLS transmitters accurately dividing the bays and accurately be utilized to provide the required patient centric identification.

While the registration is in process, a supporting nurse places the patient on a physiological monitor. The patient is associated with the patient monitor by wirelessly scanning the wristband (barcode or RFID) and completing the positive patient identification process. Within the Intelligent Hospital, physiological parameters are continuously monitored using a network-based system, providing patient device association, real-time measurement, visualization, archiving, and dissemination of critical parameters. The physiological data is available throughout the hospital, centrally on the Wall of Knowledge located in the ED, within the institutional war room, and can even be accessed from home. Connectivity to the EMR was automatically established when the medical device was associated with the patient through the positive patient identification process.

As the emergency team is completing the initial assessment, a nurse draws standard trauma labs by using a barcode or RFID-based sample collection system, scanning the patient's wristband, and printing sample vial labels directly at the bedside. Confirming the patient's identification is critical to patient safety minimizing the chance of mislabeling errors and ensuring that all the drawn samples are verified and associated with the correct patient. Once the labs are collected, the nurse uses the handheld communications device to call or notify the lab via text that the samples are ready to be delivered to the laboratory via the pneumatic tube system. This tube system also uses barcode technology to uniquely identify samples to ensure accurate delivery, save time, and limit the need to re-draw from the patient. The system enables a variety of specimens, blood samples, supplies, and pharmaceuticals to be rapidly routed and delivered to a variety of locations within the hospital.

Accurate delivery is confirmed in real time on the status board confirming lab results ordered, collected, and received by the lab. Once lab results are available and entered into the lab information system, a notification is then sent directly to the point of care and can also be viewed on the bedside PC or handheld device. In the same way, reviewed images are stored in the PACS and available at the bedside.

In the common case where patients become agitated, resulting heart rate increases and pulse oximetry decreases are indicated by the monitoring system. The nurse pushes the Staff Emergency button on the patient station for extra assistance. Activating this button generates a call at the staff station as well as notifying the appropriate supporting staff responders. Emergency requests can be visualized on the central nursing console and routed directly to the supporting staff's handheld device. At times, while trying to settle the patient, it is hard to reach the wall console. In these cases, it is possible to issue a staff emergency call from the staff worn RTLS Badge Tags. With a simple button push or pull (depending on the RTLS system), multiple staff members wearing badge RFID tags can be notified of this emergency and the user's location which appears as a text message on their badge. In conditions of worsening symptoms, such as those caused by hypoxemia, the patient may need to be intubated for ventilation management. To enhance ventilator patient management, the Intelligent Hospital has deployed real-time surveillance of ventilator patients using Wi-Fi wireless bridges on their mobile critical devices, such as the ventilator. This enables patients to be monitored anywhere in the hospital regardless of their location. All ventilator and oximetry parameters, respiratory rate and tidal volume along with pulse oximetry can be wirelessly monitored in real time at any location including the Wall of Knowledge within the ED, the central war room as well as delivered to mobile devices throughout the facility.

Patient Centric Positive Identification

In patient care areas, patients are commonly connected to a variety of devices, such as physiological monitoring, ventilators, and infusion pumps. It is vital that data from these discrete devices and systems share the information to enable an enhanced view and interpretation of the patient's status and condition. For medical device data to integrate with other applications like alarm management or the EMR, there needs to be a confirmed association of the devices with the right patient. The first step is positively identifying the patient via barcoding or wireless RFID technology. The patient bedside connectivity device provides positive patient association with connected medical devices. This process can be facilitated through wireless readers built into the bedside connectivity device and confirmed by the nurse. Patient positive identification and association is used for many applications, including connectivity, medication administration, sample collection, and patient-to-staff communications.

Patient Experience

Within the ED, patient status is continuously monitored using the RFID-coupled workflow status dashboard highlighting the state of the entire ED at a glance. On view identifies patients who are waiting to be seen by a clinical staff member (nurse or physician). As the nurse enters the bay to take vital signs, the RFID system will automatically detect the patient-staff interaction and automatically assign the nurse to the patient and update the information. When the physician consults with a patient, the RFID system detects the physician-patient interaction and removes the patient from the "Waiting for Doctor." These applications and workflows monitor and determine efficiencies of the patient care process as well as provide information to management to optimize the processes.

Medication Administration

In the case where the patient requires IV fluids or medication administered via a smart IV pump, an Intelligent Hospital deploys an advanced infusion system. The Intelligent Hospital has deployed an advanced infusion system operating wirelessly with interoperability with the EHR system. This integration helps to close the safety loop in the medication administration process by associating the pharmacy validated order with the patient, the medication, the IV pump, and the patient record using barcode medication administration technology. The nurse can use identification technologies, such as RFID, or scan barcodes on the patient ID to uniquely identify the patient, the medication, nursing staff, and the smart pump to initiate infusion based on the physician order information from the EHR to the pump. This integrated system eliminates manual programming steps and decreases opportunities for error. Running infusion status is also sent from the infusion system to the EHR to promote more timely, accurate, and complete clinical documentation in addition to enabling infusion surveillance.

Critical Devices Location and Tracking

In order to quickly locate required mobile medical equipment, such as an infusion pump, we can utilize the RFID system to locate and determine the availability of all the hospital infusion pumps. RFID enables all devices to be tracked globally throughout the institution. Since the RTLS solution typically integrates with various institutional applications, such as an inventory management system and biomed database, accurate device location and status are always available. Not only does real-time operational intelligence enable us to find the right piece of equipment for the patient – thus improving patient safety – but also knowing the location and status of our medical equipment helps save substantial dollars – right-sizing inventory levels to the optimal levels needed.

Battery Backup

A common issue within high traffic areas, such as the ED, is that battery powered devices are not always effectively managed and in a charged state. Since patients within the ED are periodically located in virtual spaces lacking proper power access, mobile devices, such as infusion pumps, can be powered by an advanced battery system, separate from any device internal battery. With this battery system, there is no downtime and never a disruption to workflow. Additionally, this single battery can be used cross-functionally with other medical devices, such as computer carts, vital signs monitors, portable EKG systems, and many other technologies. All battery data is stored in "the cloud" enabling battery management to be simple.

Patient Transfer and Preparedness

Within the typical ED scenario as patients are discharged, transferred to other hospital areas, or admitted, room-level systems such as the Nurse call system can integrate with other hospital systems such as bed management, cleaning services, and biomed automatically notifying these services to prepare for patient transfer and preparing the ED location for the next incoming patient.

Surgical Suites/Operating Room Workflow

Workflows within the surgical environment can be very different from those in other patient care areas. Similar to other patient treatment/care areas, the OR unit combines a robust network and wireless infrastructure, coupled with real-time operational intelligence from multiple RTLS/RFID systems with integrations including several modalities of communication and visualization that enable workflows that provide rapid enterprise-level interaction Within today's advanced OR suites (Figure 2.4), we find the

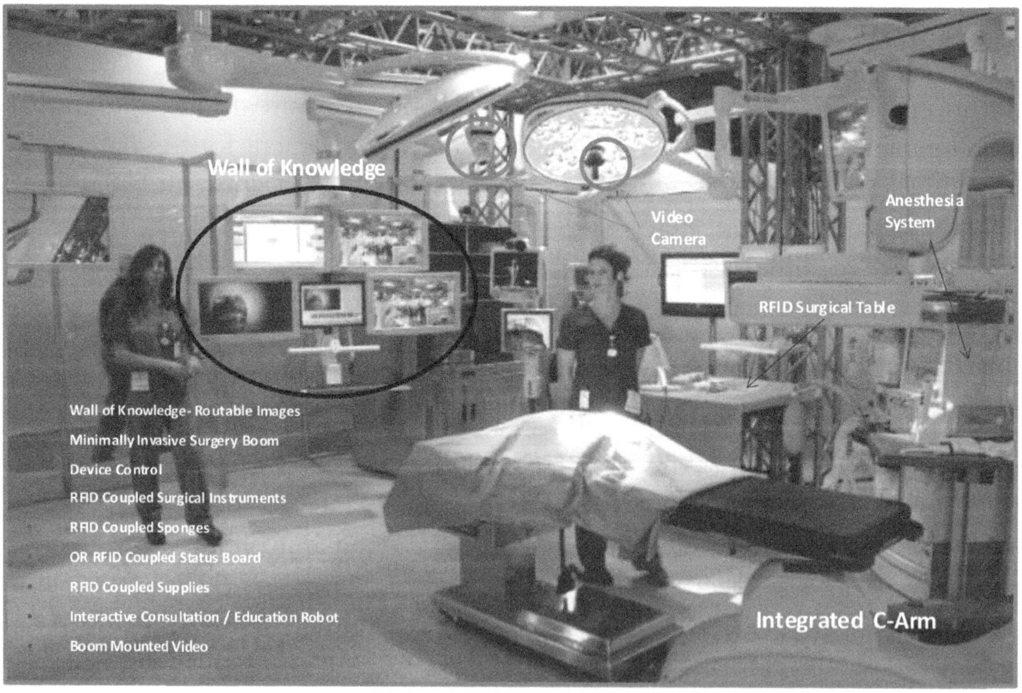

Intelligent Hospital Pavilion HIMSS

Figure 2.4 Integrated hybrid OR.

foundation of a modern OR infrastructure consisting of a hybrid surgical environment with state-of-the-art surgical tables and high intensity, advanced optics, and LED surgical lighting systems. The OR supports integrated precision HD camera systems providing multiple views of the surgical field via the boom mounted video routing and control platform. By utilizing an intuitive touch interface, our Intelligent OR medical staff can display, share, and archive digital media and many other clinical information sources. The OR uses advanced fiber optics routing, surgical video feeds, and any visual display sources to other areas of the hospital, such as the War Room, for central planning, monitoring, consulting, and education.

In a hypothetical clinical scenario, a notification has just been routed to the OR via the institutional nurse call system that the OR and surgeons are ready for the patient to be transferred directly to the OR. The Intelligent Hospital deploys multiple wireless communications delivery modalities ranging from a simple legacy pager to an advanced voice activated pendant. The advanced pendant enables hands free staff-to-staff communications to any handheld device providing comprehensive access to hospital information systems and alarm notifications. The operating room staff uses the networked displays of the hybrid OR platform or other wireless handheld devices (iPad, iPhone, tablet) to access the incoming patient's EMR information and notes to review any results available.

The OR is also equipped with an electronic monitoring system, a fire extinguisher ensuring 24×7 readiness, and Joint Commission compliance. As the threat of OR fires is high, with over 650 each year, ensuring the fire extinguishers are pressurized and clear of obstruction is vital. The fire extinguisher's obstruction monitoring system ensures the fire extinguisher alarms anytime it is blocked. The OR's supplemental oxygen tanks are also continuously monitored for readiness and pressure through an RTLS-enabled smart oxygen regulator; this tank is centrally monitored.

In preparation, the surgical technician immediately proceeds to the central sterile instrument storage area and pulls a surgical tray assembled to provide all the specific surgical instruments matched to the upcoming procedure. The tray was prepared in central processing and is equipped with active RFID tags and surgical instruments integrated with passive RFID tags enabling institutional tracking and confirmation of the tray availability within the OR.

As the patient arrives in the OR, the anesthesiologist notes that the patient needs central line venous access and has the nurse obtain the appropriate supplies including an IV pump and central line supplies.

RFID tagged equipment, such as IV pumps, is managed utilizing an RTLS system and can be located by routing the location maps to the OR monitors. As the surgical staff exits the OR, the personnel tracking system notifies us that the staff is no longer in the room, and it will automatically update as staff enters or leaves the area. returns. Similarly, as the nurse returns with the IV pump, the device is identified as present in the OR on our equipment RTLS platform interface. All devices can be identified and associated with the case procedure and recorded. The ability to track items throughout the facility results in dramatic improvements in patient throughput and equipment utilization within the hospital. The OR staff can quickly prepare and deploy the vast number of items needed for surgery because of the implementation of a robust inventory and materials management RFID solution.

These advanced infusion pumps are integrated with the Medication Administration system and the EMR. Barcode scanning or using RFID associates the infusion pump with the patient documented in the EMR. This association allows the flow of information back and forth (bidirectional) so that the pump can be automatically programmed with the correct infusion settings. This integrated infusion system also allows for automated documentation of the infusion information into the EMR, enhancing patient safety, workflow efficiency, and cost avoidance due to adverse drug events.

Utilizing an integrated passive RFID solution, tagged surgical instruments are tracked within the OR. As tagged instruments are taken off the RFID smart tray, they are registered as "not located" until they are returned to the reading device. As the technician places instruments back on the smart tray, they are restored to their located status. The surgical device tracking assures constant accuracy of surgical instruments within the OR. As a result, patient safety is improved by eliminating retained surgical item (RSI) occurrences. Similarly for surgical sponges with embedded passive tags are counted within the package at the start of the procedure. Sponges are discarded in the RFID couple container keeping track of the used sponges. Upon completion of the surgical procedures, the starting sponge count is compared to the remaining unused sponges and sponges discarded in the container. Any count variance triggers an alert which starts a patient scanning process to ensure no sponges have been left in the patient. This workflow enhances patient safety, reduces the need to reopen patients, a reportable event, and significantly reduces surgery and OR time.

The anesthesia team is supported by an advanced anesthesia delivery system, integrating the latest in physiologic and vital signs monitoring. This advanced system is interfaced with anesthesia-specific analytics software to continuously analyze anesthesia-related perioperative data which further integrates with the hospital EHR providing a comprehensive record of the procedure. The Anesthesia Information Management System (AIMS) delivers the perioperative information anesthesia providers need – including data points from the physiological monitors – to avoid redundant documentation. With unique workflow wizards and an easy-to-use touch application interface, the system facilitates pre-op and intra-op documentation to ensure provider focus remains on the patient. Utilizing the powerful interconnectivity features offered in this modern anesthesia platform, it is possible to transmit data into the EHR and War Room simultaneously.

After the patient is anesthetized and the airway secured, the patient is prepped and draped for the emergent surgery. The supporting War Room can monitor this critical situation and is in direct communication with the blood bank to facilitate the acquisition and delivery of units of cross-matched blood that have been ordered for surgery.

Surgical Information Systems

An intelligent OR View dashboard consolidates and summarizes specific patient data from electronic surgery documentation and a variety of disciplines – including anesthesia and procedural information such as medical surgical history, demographics, allergies, laboratory results, surgical status, and a timeline of the procedure. In addition, the intelligent OR View summarizes real-time data, such as current OR staff, a time-out checklist for patient safety, and physiological monitoring parameters. The information is routed to monitors throughout the surgical department, providing remote caregivers a clear view of patient and surgery status. Optimization of surgical scheduling, preadmission testing, nursing documentation, and charge capture enhances workflows and accuracy. This system also provides patient and surgical case tracking, providing visibility into specific case details and the overall status of multiple OR procedures in progress. Integrations with the OR planning system improve the operating room information flow and patient care management.

In surgical cases where consulting is necessary, for example, a mass is identified at the surgical site. A remote visualization of the surgical field coupled with direct telecommunications capability regardless of the distance enables unprecedented reach and expert consults between surgeons and specialists at distant locations. A pathologist is able to peer into the abdomen and visualize the mass with the aid of high-definition cameras and can advise accordingly (i.e., recommends a simple resection and pathology evaluation of the specimen). Specimen evaluation may require a sample to be provided to the laboratory for STAT pathologic review. Pathology results are directly delivered to the surgeon via video field OR View or to the handheld device, enabling the surgeon and pathologist to discuss and review the pathology slides.

At the start of specific phases of the surgery, timers can be set to monitor and report the status of the procedure for planning next cases or provide updated information to family members. This notification ensures compliance with family communication initiatives and results in improved patient/family satisfaction. The staff member can also send the status update to the family member through an RTLS Badge Tag assigned to the Patient's Family. Since this is a portable device, the family member can receive the status update while they move throughout the facility. This information can also be sent to the large screen in the waiting room, which automatically updates with the patient information to allow the family to see the progress.

As part of the procedure, a small fracture ortho kit has been brought from central storage to the hybrid OR, and it is determined upon opening the kit that an item is missing from the kit. This happens relatively often and is a major disruption to the surgical schedule and patient safety. The circulating nurse quickly finds the second kit using a handheld RFID reader. The kits are removed from the RFID-coupled cabinet and brought back to the operating table and the procedure is ready to start. The rapid response to exceptions such as this one is critical to patient safety. The smart cabinet solution delivers the benefit of shared visibility, efficient inventory distribution, enhanced implant tracking, and improved regulatory compliance. Utilizing RFID tagged inventory and cabinets with embedded RFID readers, simply removing items from the cabinet enables advanced inventory management and supply control.

All products and supplies used during the procedure are documented using RFID or scanned using barcodes for complete usage capture and the information is sent via interfaces to hospital core clinical documentation systems, materials management, and charge capture systems.

At the conclusion of the surgery, our wireless synchronized clock system provides precise time data throughout the entire hospital facility.

By leveraging integrated synchronized timing solutions, the OR is able to ensure billing and medico-legal accuracy and a coordinated workflow. Elapsed time records are also automatically captured documenting the procedure.

Before ending the case, embedded RFID technology revolutionizes surgical instrument management by enabling sterile processing workflow optimization required for an efficient OR. This Sterile Processing validation module ensures that all surgical instruments delivered and utilized in the OR have been accounted for and returned to the Sterile Processing Department, providing immediate alerting capabilities for any missing instruments. The SPD packing station ensures 100% accuracy of surgical sets delivered to the OR staff and eliminates case delays due to missing or wrong item delivery. OR pre-clean, wash, and sterilization stations ensure proper processing of all surgical instruments as per manufacturer instructions for use (MIFU), utilizing barcode, data matrix, and active and passive RFID technologies. A similar process if followed to ensure surgical sponge counts an accurate and confirm all sponges are accounted for.

As the patient is transferred from the OR, housekeeping is made aware of the status update within the OR suite. The OR RTLS platform can automatically notify the housekeeping staff that an OR cleaning is needed. The War Room is also able to visually confirm the current status of the OR via video feeds routed through the integrated room-monitoring camera. Once the OR is cleaned and turned over, it is noted that the OR on-call team is now available to treat other patients scheduled for surgical intervention.

Cath Lab Cardiology Suite Workflows

In the Intelligent Hospital, the Cardiology suite functions as the command or data center for Cardiac patients as well as the procedure and control suite for Catheterization and Electrophysiology procedures (Figure 2.5). The Cardiology data center enables Cardiologists to access a variety of patient centric information, including real-time physiological monitoring data, historical ECG, Holter data, stress/echo results, and Catheterization images.

Procedure Areas

Similar to all the areas of the Intelligent Hospital, the Catheterization Suite incorporates the institutional underlying high performance wireless network infrastructure enabling the new technologies and applications being

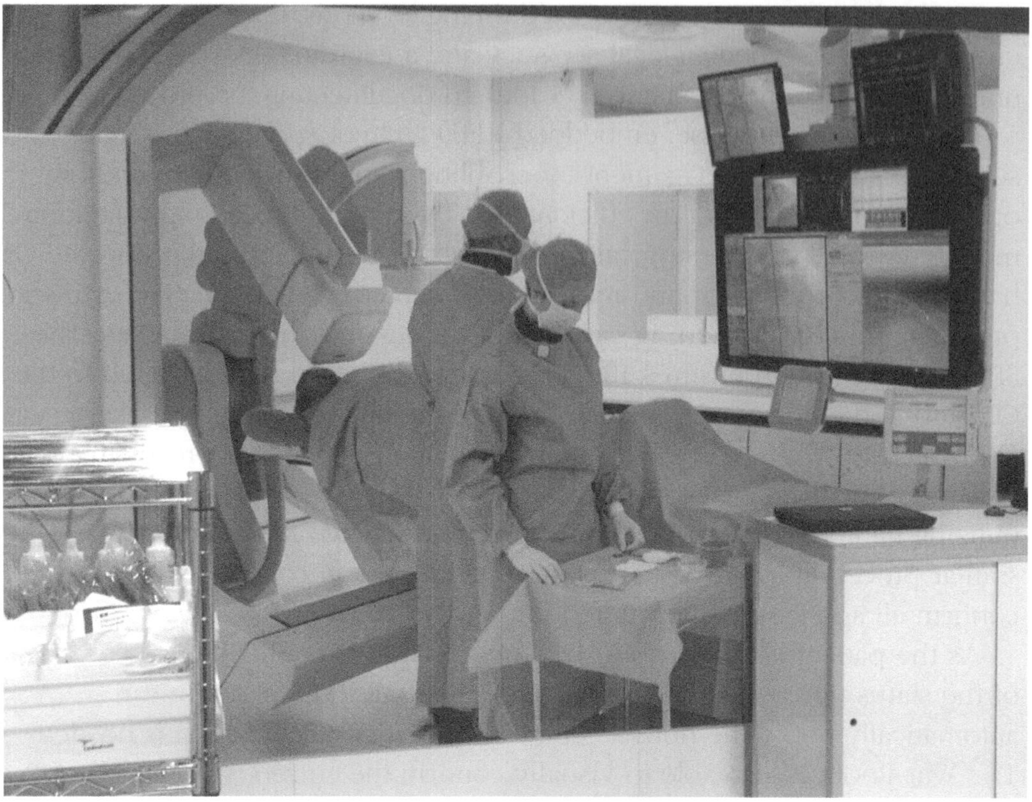

Figure 2.5 Cath Lab cardiology suite.

integrated into the Intelligent Hospital operations and workflows. Smart sensors display, log, and transmit temperature, humidity, and differential pressure data to a central software platform, generating audible and visual alerts.

Cath and EP Labs

Expanding on the ED scenario, an ED patient is en route to the ED and is transmitting his 12-lead ECG data and analysis, along with vital parameters, including invasive pressure, temperature, capnography, SpO_2, and cardiac output from the ambulance defibrillator in the EMS vehicle. The hospital infrastructure enables the ambulance ECG data to populate the Cardiology information systems and be visible for remote viewing and triage in the Cardiology Suite. In this case, the patient was triaged en route by the ED physician with consultation by an electrophysiologist bypassing the ED altogether and transported directly to the Cath laboratory.

Pharmaceuticals

As the team prepares the patient for the catheterization procedure, perioperative antibiotics are administered. Patient and pharmaceuticals are validated support through the hospital's information system and electronic medical record which is used to confirm that the correct drug is dispensed and administered, providing clinical decision support with drug-drug, drug-allergy, and five rights checking and validation.

Supplies

During the course of the procedure, a variety of supplies would normally be required. Commodity items are delivered using a 2-Bin Kanban RFID-coupled solution. This is very cost-effective solution for those items that do not need to be individually serialized. Items delivered via the 2-Bin Kanban system can be associated with a patient record for billing or traceability by using a barcode reader to scan the supply identification barcode. The 2-Bin Kanban software automatically re-orders for bin replenishment and calculates any abnormal demand variation. The software alerts the materials manager when bin quantities require adjustment.

Supply Management

Supply Chain coordinating with supply manufacturers ensures that high-value supplies are pre-tagged RFID with RFI identifiers, arrive at the hospital, and are placed directly on the RFID-enabled smart cabinets and read into the institutional inventory/asset management system. All product information is scanned from the GS1 barcode and associated with the RFID tag on the product. This eliminates the time spent tagging items at the hospital. The RFID-enabled smart cabinets provide automated inventory counts and the location of products.

At the time of the procedure, the encounter number, medical record number, procedure type, and physician and staff information are associated and entered into the usage capture station of the cabinet to ensure accurate usage capture. RFID-tagged products are waved by the RFID reader and associated with the procedure. This eliminates the need for manual recording and association of used products, as well as preventing documentation discrepancies. Inventory levels are updated, and charges are posted to the billing system based on the items that were captured during procedure time. Upon completion of the procedure, any unused products are returned to the cabinets and automatically read back into inventory without any staff intervention.

Cardiac Catheterization

"Cardiac catheterization" is a general term for a group of procedures that are performed using this method, such as coronary angiography and left ventricular angiography. These invasive procedures insert a catheter, typically through the femoral artery, directly into the chambers of the heart. Once the catheter is in place, it can be used to perform a variety of procedures including angioplasty, percutaneous coronary intervention (PCI), angiography, balloon septostomy, treatment of stenotic heart valves, closure of some congenital heart defects, placement of temporary pacemaker implantations, chamber pressure, and an Electrophysiology study or Catheter ablation.

In the case of the Cardiac Catheterization procedure tested for blood flow in the coronary arteries, blood pressure in the heart chambers, valve function, and to identify any abnormalities. A variety of interventional procedures can be performed directly in the Cath Lab or further treatments recommended. The outcome of this procedure identified that the patient exhibited both coronary artery disease and functional issues with the Aortic valve dictating the need for valve replacement.

In this scenario, the patient can be admitted directly to the Cath Lab for preparation and scheduling of the surgery as soon as possible. Catheterization data and images are immediately available within the Cardiology information systems for review, along with other archived data from other diagnostic tests and previous visits, enabling the surgical team to prepare for the upcoming surgery.

Cardiology Information Systems

Cardiology assessments and treatment plans are based on a broad base of data available and aggregated from multiple diverse systems and analytic applications, including ECG, echo stress, etc. ECG management system provides the software tools to analyze, view, edit, and compare ECG records; as well as generate, manage, and distribute reports in various formats to meet specific needs. The Information Systems enables patient centric information from multiple devices/systems to be aggregated, filtered, and processed and enables a broad view of the patient's medical status to be presented. This information can further be combined with other patient archives such as the PACS along with clinical parameters from monitoring and diagnostic devices.

Electrocardiography

Electrocardiography (ECG) is a transthoracic capture and interpretation of the electrical activity of the heart over a specific period of time, as detected by electrodes attached to the surface of the skin and recorded by a device external to the body. The recording produced by this noninvasive procedure is termed an electrocardiogram. An ECG is used to measure the heart's electrical conduction system. It picks up electrical impulses generated by the polarization and depolarization of cardiac tissue and translates them into a waveform. The waveform is then used to measure the rate and regularity of heartbeats, as well as the size and position of the chambers, the presence of any damage to the heart, and the effects of drugs or devices used to regulate the heart, such as a pacemaker.

Holter Monitoring and Analysis

A Holter monitor is a portable wearable device for continuous and prolonged monitoring of the heart's electrical activity. Electrodes are placed over bones to minimize artifacts from muscular activity. Holter monitoring is used to determine how the heart responds to normal activity and is commonly used for post-cardiac surgical monitoring or after a heart attack. The patient will wear and be monitored in his home with a Holter monitor post procedure/surgery. Heart activity will be downloaded and processed at regular intervals to enable physicians to evaluate his status and the effectiveness of his procedure during his recovery process. The analysis software processes the recorded ECG signals and determines the variations in the heartbeats and rhythms. The automatic analysis provides the physician with detailed information about heartbeat morphology, beat interval measurement, heart rate variability, rhythm overview, and patient. Advanced systems also perform spectral analysis, ischemic burden evaluation, a graph of the patient's activity, or PQ segment analysis.

Stress Testing

Cardiac stress testing provides a measure of the heart's ability to respond to external stress in a controlled clinical environment. The stress response is induced by exercise or drug stimulation. Cardiac stress tests compare the coronary circulation while the patient is at rest with the same patient's circulation observed during maximum physical exertion, showing any abnormal blood flow to the heart's muscle tissue (the myocardium). The results can be interpreted as a reflection on the general physical condition of the test patient. This test can be used to diagnose ischemic heart disease and for patient prognosis after a heart attack (myocardial infarction). The cardiac stress test is done with heart stimulation, either by exercise on a treadmill, pedaling a stationary exercise bicycle ergometer, or with intravenous pharmacological stimulation, with the patient connected to an electrocardiogram (or ECG). Patients who cannot use their legs may exercise with a bicycle-like crank that they turn with their arms.

Nuclear Stress Test

Myocardial perfusion imaging typically uses a radiotracer injected during the test procedure. After a suitable waiting period to ensure proper distribution of the radiotracer, scans are acquired with a gamma camera to capture images of the blood flow. Scans acquired before and after exercise are analyzed to assess the state of the coronary arteries of the patient by showing the relative amounts of radioisotope within the heart muscle. Stress and potential cardiac damage from exercise during the test are problems in patients with ECG abnormalities at rest or in patients with severe motor disability. Pharmacological stimulation from vasodilators, such as dipyridamole or adenosine, or positive chronotropic agents that increase the heart rate such as dobutamine can be used. Testing personnel can include a cardiac radiologist, a nuclear medicine physician, a nuclear medicine technologist, a cardiology technologist, a cardiologist, and/or a nurse.

Stress Echo/Ultrasound

An Echocardiogram Stress Test or Stress Echo is a test that combines an ultrasound study of the heart with a stress test. A stress echo looks at how the heart functions when it is made to work harder. The stress echo is identical to the stress exercise test, except, an echocardiogram is performed before and after you exercise. The stress echo is especially useful in diagnosing coronary heart disease and the presence of blockages in the coronary arteries (the vessels that supply oxygen-rich blood to the heart muscle).

Cardiology Management System

The Cardiac Management System is an integrated multi-modality image management system for cardiovascular information (Echo, Cath, CT, MR, NucMed, ECG, and EP). It provides access to multiple applications within a single flexible workspace which can be implemented in multiple locations throughout your hospital and across multiple sites. It improves clinical workflow by providing a single access point for advanced clinical applications, multi-modality images, and reports. It interfaces with advanced quantification tools within a broad spectrum of modules, suitable for multiple cardiovascular specialties including:

- **In the Echocardiography Lab** – used for viewing, quantification, and reporting of cardiovascular ultrasound.
- **In the Catheterization Lab** – it provides the tools your Cath Lab needs to enhance patient diagnosis from procedure to review to archive.
- **In Nuclear Cardiology** – it allows users to quantify gated and non-gated cardiac SPECT studies to integrate perfusion and function applications with integrated reporting, on a single screen.
- **In the Electrophysiology Lab** – it takes the complexity out of heart rhythm care by enabling review of EP recording signals and reports.

The systems comprehensive archive and database combines archived and aggregated clinical data at the physicians' fingertips, using industry standard data analysis tools, improving efficiency, outcomes, and the quality of care. The systems enable access to a broad base of cardiology information enabling consulting from home, an affiliated clinic, or anywhere in the hospital. For complex cases that need rapid assessment and precise clinician teamwork integrate enterprise visualization solutions enabling shared viewing establishing a collaborative seamless shared data environment. We can see that information centers such as this one consolidate and aggregate data from multiple sources capitalizing on the full potential of all the Cardiology scanners and systems enabling advanced visualization delivered to any location where the physicians are working. Advanced analytics are possible on multiple data sets simultaneously providing an enhanced level of diagnosis, treatment, and overall patient care and safety.

Critical Care: ICU Workflow

This hypothetical scenario within the ICU of the Intelligent Hospital highlights the typical workflows and scenarios that are enabled utilizing an integrated platform of diverse technologies integrated with the traditional medical devices and clinical applications. A patient admitted from the ED, a 32-year-old male patient who was one of the drivers in the car crash. He presented to the ED awake with multiple injuries; however, he deteriorated and now requires continued critical care including mechanical ventilatory support.

In transport, the staff was able to ensure faster transport time and oxygen tank readiness through an RTLS-enabled smart oxygen regulator. If the O_2 tank used in transport is low in pressure, a signal is sent to the asset

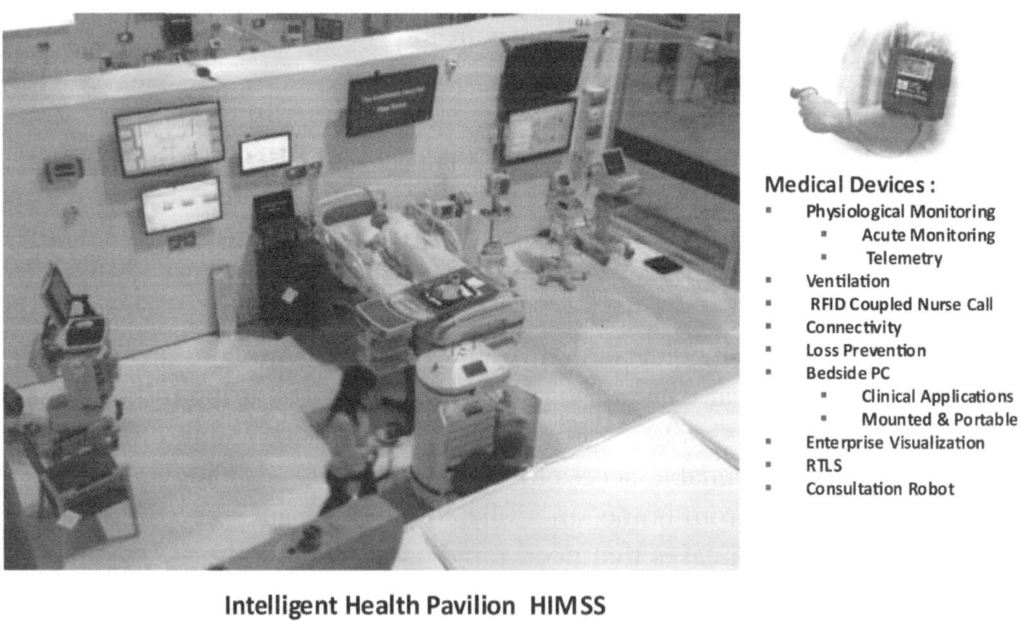

Figure 2.6 ICU/patient care room.

monitoring system, notifying appropriate staff that maintenance is needed and allowing for analytics within the Intelligent Hospital and across the healthcare system. Within an Intelligent Hospital, continuity of care has been established through medical device connectivity enabling surveillance, monitoring, and documentation, even in transport. An institutional RFID system provides the ability to locate the patient en route. This increases patient safety by allowing war room clinicians to support, monitor, and communicate, regardless of location.

This patient room (Figure 2.6) is equipped with a bedside PC, which provides access to all the hospital applications and provides the clinician bedside access to applications such as the PACS, Lab data, cardiology management, or medical history data.

As clinicians enter the ICU, they wash their hands. The staff identification RFID tag, combined with the RFID reader integrated in the dispenser, automatically identifies that the clinician washed their hands. Hand hygiene is significant for infection control; therefore, the Joint Commission Standards require hospitals to assure and measure compliance. This hand hygiene monitoring solution can also provide validation and metrics to the administration or regulatory organizations. In addition, high-touch surfaces like the infusion pole, tray table, instrument cart, and more can be replaced with Antimicrobial Copper equivalents to supplement hand hygiene and routine disinfection practices. Continuously active Antimicrobial Copper surfaces can reduce bacterial loading by greater than 58% during active care, which has been proven to reduce the rate of healthcare associated infections.

Typical of most high acuity patient environments, our patients lie in an instrumented smart bed giving a secure dynamic environment capable of monitoring patient weight, side rail status, wheels, and patient communications. The bed integrates with a nurse call platform, providing patient-to-staff communications. Caregivers and staff wear a staff badge that uses IR, ultrasound, or low frequency RF technology through receivers located throughout the hospital and patient rooms. Calls and alerts such as bed exit or rails down can be immediately directed to the appropriate and available caregiver through the staff worn or handheld wireless device. In this example, the rail down alert is sent to the primary caregiver which enables the care provider to respond via her handheld device that she is unavailable to respond which in turn triggers the alert to be sent to the secondary caregiver who acknowledges and accepts the alert and attends to the potential bed exit alarm.

Physiological Monitoring

Within the ICU, the patient is placed on physiological monitoring as ordered by the physician. Within the Intelligent Hospital, physiological parameters are monitored by a high acuity networked monitoring platform. This provides real-time measurement, visualization, archiving, and dissemination of critical parameters. Critical parameters such as heart rate, ECG rhythms, blood pressure, and SpO_2 are continuously monitored and are also available in the War Room and on mobile devices carried by nurses and doctors. At any time, the nurse can initiate a 12-lead ECG for archiving in the hospital Cardiology management system.

Vital to the care of this patient is BIS monitoring, or Bispectral index to measure the depth of sedation. The BIS Unit is missing, but through an RFID-coupled medical device database, the nurse can confirm that there are devices available within the ICU. These devices do not have active RTLS tags due to cost and/or size constraints and are tagged with a small passive RFID tag. The staff uses a handheld RFID scanner to quickly scan the local area, locate the BIS module in the storage closet, and connect it to the physiological monitoring platform.

All the monitoring parameters can be distributed to multiple locations, including other units, patient rooms, a physician's office, home offices, handheld devices, alarm management systems, and the EMR.

The Intelligent Hospital utilizes a combined approach. This includes centralized monitoring, which delivers information to a war room for continuous surveillance, and a distributed model where critical data and alarms are simultaneously being delivered to the point of care via handheld devices carried by the staff. Also delivered to the point of care are any critical test results from departments such as the lab or Radiology. This eliminates the lab tech's need to play phone tag with the attending physician when a critical result is available and automates the delivery of this information directly to the appropriate caregiver.

High-fidelity Data Collection and Analytics

To provide local and remote real-time surveillance, a multiport bridge provides high-speed, high-fidelity data acquisition from medical devices such as the described physiological monitors, ventilators, infusion pumps, etc. Continuous acquisition of ICU monitored parameters and video is acquired over the institution's network. Within the War Room, this information can be viewed and reviewed, acted upon, and distributed to third-party applications, such as EMRs or clinical archives. Real-time data may be used for active patient monitoring and alarm management or fused with EMR data elements to implement algorithms, including third-party applications, analytics of clinical decision support to identify and prevent adverse clinical events.

Additional functionality may include the integration of activity or temperature monitoring, using a wireless temperature transducer and/or accelerometer directly to the patient, providing continuous patient body surface temperature and archiving this information for display over the wireless network.

The ICU room can also be configured with a video camera enabling remote clinical observation. Advanced configurations can include two-way audio and visual communications between remote intensivists, hospitalists, and in-room caregivers.

Connectivity

In this acute care setting, patients typically are connected to a variety of other devices, such as infusion pumps. It is vital that data from these discrete devices and systems share the information to enable an enhanced view and interpretation of the patient's status and condition. For medical device data to integrate with other applications like alarm management or the EMR, there needs to be a confirmed association of the devices with the right patient.

The first step is positively identifying the patient through a barcode or, ideally, an RFID wristband. The patient bedside connectivity device has an RFID reader that automatically detects the presence of the patient and the presence of the medical devices, thus providing positive patient association with connected medical devices. This process is automated using passive RFID built into the bedside connectivity device, and the nurse confirms the device association to the patient on the touch screen.

On a unit central station or dashboard or room display, clinical staff or care providers can monitor the status of the device changes from available to unavailable as the pump is positively associated with the patient, therefore unavailable for use with other patients. This automatically collected data can be aggregated across the intelligent Hospital, and across the entire healthcare system, to determine enterprise-wide utilization rates and optimal inventory levels.

The nurse can receive secure messages over her VoIP device/phone, including messages, such as notifications that the intensivist will be coming to the unit to insert a central line for hemodynamic monitoring. Each staff member within the Intelligent Health System carries a VoIP device that enables direct staff-to-staff communications. The system integrates with the hospital phone systems and additionally enables incoming and outgoing calls directly to and from the care provider.

Hospital procedures and protocol dictate that the central line kit is set up and ready when the physician arrives. A nurse places the kit on a procedure tray, alongside the other supplies required to perform the procedure. In order to ensure supply availability and demand accuracy for this procedure, inventory is managed using the two-bin RFID practice. This practice eliminates the need for manual demand assessment rounds by triggering replenishment requests by users when a predefined batch of a product has been depleted. Only when a bin becomes empty will the user transfer the RFID-enabled location tag to the replenishment board. A secondary batch of the same product holds sufficient stock to support demand during the replenishment cycle. Materials management brings the supplies requisitioned and rotates the remaining stock while returning the RFID-enabled tag to its location to reduce the risk of product expiration. Unitary consumption can also be captured and linked to the patient's case file.

For rapid assistance with patients becoming agitated and difficult to manage, a nurse may find it necessary to contact the attending physician or other staff to assist either via handheld or RFID emergency button, through the attending physician button on the workflow. The physician responds directly to the nurse's handheld device and in response orders the immediate administration of 4 mg of morphine and 5 mg of midazolam IV, followed by a midazolam drip at 2 mg/hour.

Medication Administration

The nurse goes to the automated dispensing cabinet to withdraw the medications. The logon screen displays information such as drawer temperatures and alerts on any expiring or expired medication. The nurse logs in and selects the Dispense Medication by Patient option. Identifying and selecting the patient, in this instance. From the list of all physician ordered medications for this particular patient, the nurse selects the required medication for dispensing, after which the drawer opens, and the correct compartment is identified on the screen. The nurse removes the drug from the indicated compartment, closes the drawer, and awaits clearance to proceed. Once confirmation for the removal of the correct drug is provided, the system inventories itself, eliminating item-level scanning and manual counting. In the unlikely event that the wrong drug is removed, the drawer will reopen for the nurse to correct the selection.

Staff's final opportunity to ensure medication safety is at the point of administration. Prior to every medication administration, the nurse places the medication on the tray, scans the patient's RFID identification wristband to positively identify the patient, and associate the patient with the medication on the tray. In the case of a serious medication error where the nurse inadvertently selects the incorrect item for this patient, an alert is automatically triggered averting this error condition. The nurse replaces the item with the correct one and rescans and confirmation is provided. This time, a correct medication confirmation is received on the screen as defined within the physician's order. Once the verification process is complete, the medication can be administered safely to the patient.

To administer specific medications, commonly an infusion pump is required. Quickly, the nurse goes to grab the nearest pump to administer the medication using RFID technology to present device status. The nurse can tell that the nearest pump is identified as dirty. To save time, the RTLS system could have been used to locate the nearest available clean pump, as indicated. Each device is fitted with an RTLS tag that enables the institutional tracking and location of possible replacement devices. This real-time map is used to locate all clean and available pumps, as illustrated. In this case, an available clean pump is across the room.

In addition, since Wi-Fi RTLS systems can track Wi-Fi devices as well as tags, you can do location-aware searches. Clinicians can use "Find near me" features on mobile devices to find the nearest available pump to their specific location. If no pumps were locally available, the nurse can submit a request through the Asset Management system. This sends a request to Central Supply and simultaneously to the technician, generating equipment orders for delivery of the asset to the patient room. If a pump needed preventative maintenance, the device and location are recorded in the biomedical engineering database, initiating work orders for repair. Central Supply is automatically notified that the number of available pumps in the ICU has dropped below the minimum level, and a replenishment service request is dispatched to the closest worker. Not only does a real-time operational intelligence system enable us to find the right piece of equipment for the patient improving safety, but knowing the location and status of our medical equipment helps save substantial dollars by right-sizing inventory to optimal levels.

These advanced infusion pumps are integrated with the EHR, which creates an integrated system connecting the infusion administration workflow to the pharmacy validated order. By adding the barcode scanning of the infusion pump, you create an association between the patient, the pump, and the EHR. This association allows the flow of information back and forth so that the pump can be automatically programmed with the correct infusion settings – directly from the pharmacy validated order, which eliminates manual programming steps and the opportunity for error. This integrated infusion system also allows for automated documentation of the infusion information into the EHR, enhancing patient safety, workflow efficiency, and cost avoidance due to adverse drug events.

Commonly, medical devices, transducers, etc. become tangled with the bedding or are placed on food carts, inadvertently misplaced, or discarded. In the Intelligent Hospital, where device incorporate an RFID tag, and the laundry chute houses, and door entrances support an RFID transmitter – receiver creating a choke point that identifies and tracks the inventory passing through the chute which prevents valuable assets from accidentally being discarded by triggering various notifications or alarms. This aids in minimizing device loss and rapid recovery.

The ICU represents a high paced high acuity environment where patient outcome is a direct function of aggregated clinical information from a variety of medical devices and clinical systems from a variety of locations with the Intelligent hospital. Decision support applications that collect and process vital information enable quick decisions and actions to be performed. Many diverse technologies combine to validate information and minimize potential errors, as well as optimize the processes and workflow to provide quick response and patient safety, all of which positively impact patient satisfaction.

Patient Care Management (Patient Rooms) Workflow

Step Down, Patient Care and Ambulatory Care

As in all patient care areas of the Intelligent Hospital, a robust network and wireless infrastructure, coupled with real operational Intelligence from multiple RTLS/RFID systems with integrations including several modalities

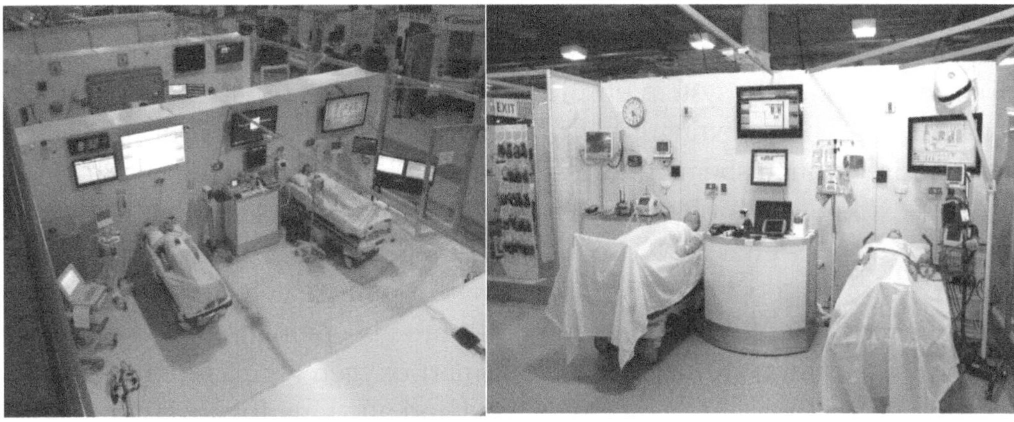

- Time Sync. Clocks
- Medical Devices & Connectivity (Barcode & RFID)
- RFID Hand Hygiene Compliance
- RFID Coupled Patient / Staff Tracking
- Patient Identification
- Label Scanners & Printers: Wristband, Sample Collection
- RFID Coupled Bed / Patient Management System

Figure 2.7 General patient care room.

of communication and visualization that provide rapid enterprise-level interaction (Figure 2.7). In addition, all clinicians carry VoIP devices, enabling staff-to-staff communications with integration to the hospital phone system, supporting incoming and outgoing calls.

It is important to note that in areas other than the acute patient setting the nursing staff to patient ratio decreases and streamlined communication improves patient care and throughput, making every minute count. The ability to quickly and easily communicate puts patients back at the center of care while improving efficiencies throughout the care process.

Vital to providing effective patient care is the management of resources, supplies, devices, and staff. Embedded RTLS solutions enable all devices to be tracked globally throughout the institution. At any moment in time, the equipment management system can use this location information to ensure that lean workflow is being followed, to reduce the critical time wasted when finding needed devices. RFID analytics report on equipment and device movements, enabling discussions on device quantity, distribution, and optimal par levels. Equipment and supplies can be automatically re-ordered from the floor when inventories fall below defined thresholds.

Throughout the hospital, these patient care areas ensure unit facility and hospital protection utilizing fire extinguishers that are equipped with an electronic monitoring system ensuring 24×7 readiness and Joint Commission compliance. The fire extinguisher monitoring technology is ensuring all hospital fire extinguishers are pressurized, in place, and free of obstruction. The fire extinguisher's obstruction monitoring system ensures that if the fire extinguisher signal is blocked at any time or a fire extinguisher is removed for use, a "CODE RED" signal is sent immediately to alert hospital staff. Similarly, supplemental oxygen tanks are also monitored for readiness and pressure through an RTLS-enabled smart oxygen regulator, notifying appropriate staff if maintenance is needed and allowing for analytics within the Intelligent Health System.

In this fictitious but very realistic scenario, two patients were just admitted, a female and her daughter, transported from the ED. They are victims of the collision and are admitted for observation. The mother is a 29-year-old female, and the baby is 3 months old. They presented to the ED awake with multiple minor injuries like bruises and abrasions. The mother suffered a concussion and is also a diabetic. Since data can be seamlessly shared in the hospital, accessing and reviewing clinical information from the assessments made during ambulance transit and in the ED, including **imaging**, as well as diabetic history from their primary physician, enables the nurses to quickly develop a treatment plan for the mother. A streamlined communication methodology is critical for efficient patient care – with a single tap on a button, the mother's and daughter's assessment can be securely shared with the family's primary physician. Based on her condition, the nurse classifies her as a potential fall risk, requiring additional observation. This is indicated with the nurse call system dome light being illuminated in yellow.

Remote Monitoring, Connectivity, RFID in Transport

In this facility, continuity of care has been simplified through medical device connectivity for surveillance, monitoring, and documentation in the EMR, even during transport. In addition, the ability to locate the patient en route immediately through an RFID tag on the patient's wrist increases safety as the war room clinicians can support monitoring and communication regardless of location.

This configuration represents a moderate acuity patient care area but is enabled to increase the patient acuity in cases of overflow and emergencies such as the COVID pandemic. This room is equipped with a bedside PC, which can be either wall mounted or in this case a handheld tablet commonly replacing the computer on wheels implementations. This bedside PC can run all hospital applications and provides the clinician direct access to the PACS, Lab data, cardiology management, or RTLS.

Patient Bed

The patient lies in an instrumented smart bed providing a secure dynamic environment capable of monitoring patient weight, as well as wheels, side rail status, and patient communications. A pressure sensitive pad is integrated with the bed to ensure patient safety. The pad issues alerts if the patient tries to exit the bed unassisted. The monitor integrated with the bed and pressure pad can be activated to detect movement that could lead to a fall from the bed.

The bed and pad integrate with a nurse call platform, providing one mode of alarm notification as well as patient-to-staff communications. The nurse call system can annunciate alerts from the bed, including bed exit alarms, rails down, or brake off alarms.

Hand Hygiene

As you can see, the clinician enters the patient, the RFID tag combined with the RFID reader integrated in the dispenser, automatically identifies that the clinician washed their hands. Hand hygiene is significant for infection control; therefore, regulatory compliance as defined by the Joint Commission Standards requires hospitals to assure and measure hand hygiene compliance. This hand hygiene monitoring solution can also provide validation and metrics to the administration or regulatory organizations. The Intelligent Hospital has additionally replaced many surfaces, such as instrument carts, gurney railings, chair arms, tray tables, and IV poles, with the Antimicrobial Copper equivalents. Research has documented that surfaces made from these materials can reduce bacterial loading by greater than 80% during active care which has been proven to reduce the rate of healthcare-associated infections.

Once in the room, the nurse confirms patient identification, using the RFID or barcode wristband, and using the bedside medication tracking system checks physician medication orders. The mobile system streamlines medication tracking and transport by communicating seamlessly with leading Automated Dispensing Cabinets. Clinicians capture patient information from these dispensing systems, eliminating redundant data entry, allowing them to stock their system in advance with the patient-specific ordered medications. The integrated medication label printer and barcode reader enables the administration of patient-specific medications to comply with the eight rights of patient administration at the bedside. If other medications are required, they can be rapidly prepared and delivered to the bedside. The physician order also requested that the patient be placed on telemetry.

Telemetry

Traditionally, patients in a Step-Down Unit only have vital signs spot-checked at 4-hour intervals. This leaves these patients at risk for complications of respiratory failure, hemorrhage, fall risks associated with trauma, and weakness. In the Intelligent Hospital, physiological parameters are continuously monitored using a continuous monitoring system, replacing the traditional spot-check monitoring. The clinician applies the telemetry unit to wirelessly transmit to a central location, commonly a unit base central station, and additionally to the institution's central War Room. Key physiological parameters, including respiration rate, oxygen saturation, heart rate, rhythm, blood pressure, and end-tidal CO_2, are recorded. The system automates the vital sign collection and clinicians are alerted if the patient's condition starts to deteriorate. These alerts can be sent to a variety of mobile devices resulting in quicker intervention to prevent potential complications. Graphical trends and detected alarms are stored for 72 hours and are integrated into the hospital EMR. Telemetry enables global defensive monitoring of any patient, anywhere within the institution. Using wireless connectivity, the clinician can view the current and historical monitoring data on any wireless device in the patient's room or anywhere inside or out of the hospital.

Infant Security

Infant security is a critical aspect of patient safety. Hospital accreditation standards define patient abduction or discharge to the wrong family as a Never Event. To ensure infant safety, the clinician attaches a small active RFID security bracelet to the baby's ankle which is associated with the mother and/or clinical staff. Any unauthorized attempt to remove the tag or to take the infant off the unit will generate an alarm at multiple locations, workstations, as well as on staff voice communication badges and can additionally lock down the area automatically locking entrances and exits.

The infant's mother also wears an active RFID bracelet, which is matched with Baby. If a nurse accidentally brings the wrong infant to the mother, the nurse is immediately warned of the mistake to help reduce the risk of nursing the wrong infant or discharging to the wrong family. All transactions are recorded in the database so that the hospital has a permanent record showing the incident and the quick response.

Medication Administration

As the nurse continues to get the patient settled, hospital protocol recommends the patient be placed on a normal saline drip. The physician has ordered a change in the prescribed pain medication to be delivered to the bedside. These medication orders were sent to the pharmacy and are being prepared. Once prepared, they will be delivered via an autonomous mobile robot with RFID tracking and biometric security. In addition, the nurse can check the status of the medication delivery on the tracking screen, helping reduce missing medications and unnecessary re-fills.

Find IV Pump

Instead of wasting time searching for an IV pump, the nurse uses an RTLS map to locate available devices. The system identifies pumps that are in use or dirty as well as those available for use. The nearest clean and available pump can be easily located and retrieved. In this case, the closest pump is marked as unavailable, needs maintenance. Fortunately, there is a clean pump locally available; however, if this pump is not available, the nurse can submit a request through central supply. The equipment management system will receive this urgent notification from the EMR that an IV Pump is needed for the patient. The order is dispatched and escalated to an available equipment handler, who picks up an available pump from a nearby clean storage area, delivers it to the room prior to the patient's arrival, and closes the request.

The system streamlines this process by identifying pumps that are in use or dirty as well as those available for use. If the nurse needs additional equipment, she can submit a request through her smartphone, which sends a request to Central Supply and generates an equipment order for delivery to the patient room. Not only does RTLS and the equipment management system support finding the right piece of equipment for the patient, improving safety, but also knowing the location and status of our medical equipment helps save substantial dollars by right-sizing inventory to optimal PAR levels.

In the Intelligent Hospital, it is the policy to administer all IV fluids using smart infusion pumps. The hospital has deployed an advanced infusion system operating wirelessly with interoperability with the EHR system. This integration helps to close the safety loop in the medication administration process by associating the pharmacy validated order with the patient, the medication, the IV pump, and the patient record using barcode medication administration technology. The nurse can scan barcodes or read passive RFID on the patient ID, the medication, and the smart pump to initiate the population of order information from the EHR to the pump. With this integrated system, the ordered infusion settings automatically populate on the pump screen, eliminating manual programming steps and decreasing opportunities for error. Running infusion status is also sent from the infusion system to the EHR to promote more timely, accurate, and complete clinical documentation in addition to enabling infusion surveillance through the EHR system.

Medication Robotic Delivery

The hospital used a mobile robot to support the medication delivery process. The robot traveled autonomously from the basement of the hospital where the pharmacy is located, operated the elevators, traveled the corridors safely navigating around people and unexpected obstacles, and arrived at the nursing unit. It announces its arrival with a voice, and the nurse securely logs in with her personal PIN, uses her fingerprint to validate her identity, and then removes the medications. The tracking screen indicates the medication has been delivered and confirms the staff member who received it. The use of robotic delivery systems requires certain pathways to remain clear. High accuracy RFID systems with resolutions of up to 18 inches or less are used to ensure devices or other equipment are not inadvertently left in the pathway. A placement of a multi array locator in key locations provides the capability to monitor these locations to the accuracies needed.

Nurse Call Pump Integration

Before setting up the pump, the nurse plugs the cable from the IV pump into the nurse call system. This allows a secondary annunciation of pump alarms to go directly to the nurse call system, which then directs the alarm to the assigned staff. This can also trigger alerts to the handheld devices and staff phones. Rapid intervention to IV pump alarms, whether due to infusion completion or interruption, decreases the potential for adverse events, increasing patient safety. When the pump goes into alarm, it is indicated with the dome light, staff console, and the staff communication badge.

Draw Labs

As a follow-up, the physician has ordered a hemoglobin and hematocrit lab. The hospital uses a barcode-based sample collection system, scanning the patient's wristband and printing sample vial labels directly at the bedside. This ensures that all the drawn samples are verified and associated with the correct patient minimizing the chance of mislabeling errors. Once the labs are collected, the samples can also be retrieved by the mobile robot. The nurse logs in with her credentials and fingerprint and loads it with the sample. She then sends it on its way with a single button push to automatically deliver the samples securely to the laboratory. Other applications of mobile robots include the delivery of meals, exchanging linens, and trash removal.

Patient Interaction and Satisfaction

To ensure a high degree of patient satisfaction, the use of an interactive patient system enables the patient to view her personalized electronic whiteboard. Through an integration with the hospital's Admissions and Discharge (ADT) system, the patient has access to a snapshot of her care team, pain management tools, orientation videos on safety, fall prevention, discharge planning, and condition-specific education based on physician orders, as well as other resources necessary to help orient the patient during their stay. Through integration with the hospital's EMR system, viewing of patient education is automatically documented into the patient's health record. The system makes Samantha an active participant engaged in her care, empowering and connecting her to her care team.

To improve efficiency and enhance the patient experience, the hospital uses an integrated web-based mobile meal ordering application accessible from the web browser on the mobile tablet or bedside device to support staff-assisted entry of patient and guest meal selections for either on-demand room service or a tray line meal service. The application interfaces with the hospital's CPOE providing real-time diet order and allergy information along with clinical tools to ensure that all foods served are appropriate and meet the patient's needs. To help ensure patient safety, the staff member is automatically alerted that the patient is a diabetic who requires insulin and that she is allergic to peanuts. The system then automatically tracks and displays carbohydrate totals compared to her prescribed diet goals reinforcing the hospital's patient education initiatives. Once the meal order is placed, tray tracking information automatically displays on monitors at the nursing unit or War Room to facilitate communication with nursing for a pre-meal blood glucose check. When the tray arrives on the floor, foodservice staff deliver the tray and update the tray tracking system using the smartphone or handheld to confirm the delivery.

Patient Communications and Alert Notification

Patient assistance is a common request in cases where patients need support to go to the restroom. By pressing the staff call button, a call is then sent to the staff worn communication badge, rings at the staff console, and lights the dome light. However, if before the staff can respond the patient attempts to go to the bathroom, it triggers a bed exit alert to the central station and to the staff phone. The pressure sensitive pad on the bed will also send an alert to other hospital systems, such as nurse call, staff worn RFID badge, or a message to the staff IoT device. The staff responds immediately to prevent the patient from falling. If additional assistance is needed, the nurse can press the RFID badge button to request assistance. This activates a display on the handheld device, identifying the person requiring assistance and their specific location. The nurse RFID tag confirms her presence in the room through an RFID–Nurse Call integration. The nurse system automatically cancels the bed exit alarm and corresponding indicators.

The nurse confirms the patient is at risk of a fall and requires assistance in getting the patient back into bed and stabilized. The nurse can issue a request for assistance on the nurse call system. The nurse activates the fall risk notification on the dome light indicating to all staff on the unit that this patient is a fall risk.

Patient Discharge Processes

As the mother and the baby prepare for discharge, the mother uses the Interactive Patient Systems accessing the module which explains the discharge process, sets expectations, and informs the patient of the ability to fill a prescription by way of the on-site retail pharmacy.

Environmental Services

The nurse removes the telemetry pack and IV access so that the patient can get dressed. Once they have left the room, the nurse takes the laundry and pushes the button on the nurse call for discharge. This sends a message to Environmental Services to request that the room be cleaned. Room status changes – from soiled, available, and occupied – are indicated on the map with a visual color change.

Equipment Shrinkage

A common occurrence during clean-up processes, portable medical devices, or otherwise valuable inventory is misplaced or discarded. In the Intelligent Hospital, devices incorporate an RFID tag, and the laundry chute houses an RFID transmitter–receiver. This creates a choke point that identifies inventory passing through the chute, preventing valuable assets from accidentally being discarded by triggering various notification modalities. If the clinician places the telemetry box on the bed, it can be inadvertently discarded with the sheets down the laundry chute. The nurse call station receives an alert from the RFID system. This choke point level of device tracking enables improved loss management and inventory shrinkage.

Pharmacy Workflows

The Pharmacy department is made up of pharmacists and pharmacy technicians who handle and coordinate all aspects of medication therapy. In most of the processes within the Pharmacy, technology is being utilized

Figure 2.8 Intelligent pharmacy.

to drive efficiency and improve accuracy (Figure 2.8). The patient critical processes required to ensure that the five rights of medication safety – the right medication, to the right patient, at the right time, via the right route, and at the right dose – are validated through workflows utilizing enabling technologies to ensure patient safety and minimize errors.

Computerized Physician Order Entry

The pharmacist has received a new urgent order via the computerized electronic order entry system. Just seconds ago, this order was placed by an emergency room physician taking care of a patient who arrived at the ED a few minutes ago with chest pain and an elevated "ST" segment on his electrocardiogram. Suspecting a heart attack, the physician has ordered a "**clot buster**" drug that will be given by intravenous infusion.

The pharmacist promptly reviews the patient profile and history within the EMR to ensure the order is appropriately entered and there are no drug interactions or contraindications. Having the full patient medical record at the fingertips of the pharmacist is a great benefit and time-saver that helps clinicians make informed decisions about the care this patient will receive. Once the order for the clot busting drug is verified by the pharmacist, a label is immediately generated from this wireless battery-enabled portable printer.

Medication Inventory

The pharmacist utilizes his proximity card to obtain the medication from this smart storage cabinet. This smart cabinet tracks, monitors, and records critical high-value inventory as it is dispensed. Each medication inside the smart cabinet has an RFID tag that contains important information about each item. When the door closes, each item is automatically read and identified against a database. In addition, the temperature of each cabinet is automatically logged, maintaining a temperature pedigree for each medication. Inventory management systems ensure inventory levels are optimized and readily available for dispensing to the patients. Current inventory information is available from any computer or mobile device. When an item drops below the minimum level, a replenishment order is automatically sent to the distributor, preventing the risk of a stock outage. Upon receiving a new delivery, the pharmacy tech simply places the RFID tagged medication inside the cabinet to reconcile the order. With real-time visibility and actionable data, this system prevents the risk of over purchasing, a costly and common occurrence. The system also tracks the expiration dates and lot numbers of all items in the inventory. Alerts are sent before an item expires, thereby avoiding loss of an expensive item, and in the event of a recall, allowing for a prompt response by pharmacy personnel.

This inventory management system makes it possible for a distributor to consign inventory to the hospital and charge them only after the medication has been used, eliminating labor intensive manual counting, reconciliation, and reporting by pharmacy staff, while ensuring inventory accountability for the vendor. Since some medications are extremely expensive and used infrequently, a consignment program enables the institution to optimize its cash flow and use of available funds.

Tube System

Once the drug has been labeled, checked, and verified by the pharmacist, it is ready for delivery to the patient care unit. In this case where the medication doesn't require any additional mixing or preparation, it will be delivered to the ED via the tube system. The Pharmacy utilizes a "secure tracked send" technique, where the pharmacist scans the drug item, scans the carrier, and finally scans their identification badge. This quick scanning process enables the Pharmacy to track the delivery and receipt of the medication. After selecting the appropriate destination from the quick pick list, the carrier will be sent and delivered to the ED. Upon arrival, the receiving nurse will also scan their identification badge to release the carrier from the system. This process ensures chain of custody is maintained requiring each step of this transport to be logged in real-time on a web-based database available at all locations. In the rare event that a medication is misplaced or there is a delay in treatment, all steps in the transport chain are documented and logged.

Infusion Using a Smart Infusion Pump

Once the nurse receives this "clot buster" intravenous medication, the nurse prepares to start the patient infusion process using a smart pump. The smart pump has an electronic brain that contains a library of drugs with individual maximum and minimum dosing limits and safety parameters defined for specific patients and care areas. These limits help prevent programming errors which can lead to over- and underdosing of the medication. Within the Intelligent Hospital, the Wi-Fi wireless system enables the pump to communicate with the hospital's electronic health record and other applications including RTLS. Once the nurse scans the barcodes or reads the identifying RFID tags, on the infusion pump and on the patient, a patient centric association is established which can then validate the physician order for the "clot buster" drug. Once the drug is confirmed, the exact dose and rate that the physician ordered can be set on the infusion pump. The nurse simply verifies the order, eliminating the need for manual programming, decreasing the opportunity for human error.

Infusion Tracking Dashboard

Once the infusion begins, this IV pump will communicate wirelessly to a central server which updates the electronic health record with visibility on the institutional dashboard. From this web-based dashboard, the pharmacist can review infusion information for all patients in the Intelligent Hospital in near real-time. The infusion pump is also integrated with the RTLS server, which continuously downloads the infusion status. Within the RTLS system, the infusion is identified as in use and no longer available for other staffing searching for an available infusion pump. This data can be aggregated and provide the institution via the dashboard the unit or institutional utilization of specific devices in this the infusion pump.

Emergency Kit and Tray Replenishment

Upon the arrival of patients into the ED, an emergency medication tray is assigned to the patient. This tray contains various medications commonly used during emergencies. There can be hundreds of these kits throughout the hospital, each designed to treat a specific emergency or situation. Managing all these kits is a challenge and can be a significant patient safety issue. The pharmacy must ensure that the contents of all trays are accurate and that there are no expired or recalled drugs contained in any trays. To ensure the highest level of patient safety, the Intelligent Hospital Pharmacy uses an automated RFID Kit and Tray Management System to help manage the process.

Placing a used tray from the ED into the RFID enclosure, the pharmacy technician presses the Scan button, and the system quickly reads what items are missing, expired, and soon-to-expire. This is based on a simple comparison of the RFID identified contents at the time of kit creation to the present state of the kit. Missing items are removed from the smart cabinet and restocked into the kit. The technician can quickly perform a follow-up scan providing a validation and confirmation scan. A report displaying the complete contents, expiration dates, and the first drug to expire is printed and then placed inside the kit. All events and data are captured in a database that pharmacy staff can query. If any medication expires, is soon-to-expire, or is recalled, an alert is generated to the staff.

Identifying products and technology that ensure patient safety is critical in all hospitals. This system ensures all kits are deployed with the right medications, in the right doses and quantities. Utilizing a system like this removes potential opportunities for error and patient harm.

Chemo Ordering and Production

The pharmacist utilizes the computer workstation and begins to review orders that have recently been entered by physicians throughout the hospital.

In this case, a new order for a chemotherapy agent has just arrived from a pediatric oncology ward for a two-year-old girl on her last round of chemotherapy. After performing several clinical checks to ensure the chemo dose is appropriate for this little girl, the pharmacist verifies the physician's order, which then sends the details into a Compounding Medication Workflow Management System within the dedicated to sterile compounding of intravenous medications.

The person working in this area is typically a Pharmacy Technician. The order sent has been received by the system which will manage all aspects of the sterile compounding process. The technician will review the order work and prioritize orders based on urgency. The process of compounding the chemotherapy begins by acknowledging the queued orders. The technician will then be presented with a picklist on which medications and supplies will be listed that the tech will gather from the storage area.

The Compounding Medication Workflow Management System manages and streamlines the process at each step along the way, offering efficient and safe preparation aids controlled by an electronic scale and integrated barcode support to decrease operator risk and improve patient safety during the production cycle. After the technician has gathered the ingredients intended for this patient, the tech will compound the medication. Scanning the first ingredient, the technician validates that the ingredient is the correct product. An alert is generated; in this case, a similarly sounding and look-alike item was retrieved in error. The technician will now scan the correct item, and the system recognizes it as the correct item and the error is averted.

The technician will now follow a standardized and automated process aided by on-screen step-by-step instructions at each step. Such aids help ensure that every technician follows the same established quality procedure and processes at each stage of compounding.

The technician performs the compounding using a special closed-system transfer device that is leakproof and airtight to prevent drug exposure to the personnel involved in the preparation and administration of hazardous drugs. Clinical studies have documented health risks of exposure to chemotherapy agents, even at low levels. This type of system is designed to prevent harmful exposure.

A closed transfer system helps improve operating efficiencies minimizing drug waste. Normally, when a vial is punctured by a traditional syringe and needle, the unused contents of the vial must be discarded in a rather short period of time because of the risk of potential microbial contamination. When using a closed system transfer device, that potential is eliminated and the unused contents can be saved for a longer period, thereby reducing potential waste. A growing number of institutions that use this type of system can not only reduce drug waste but also maximize the utilization of drugs in short supply.

In the process of compounding this medication, an error has occurred, and the technician has received a warning because the volume of the drug withdrawn from the vial is not within the tolerance limit. On-screen instructions will now aid the technician on how to remediate any preparation errors and, more importantly, will not allow the technician to proceed to the next step until the error has been rectified. Integrated quality assurance checks during the compounding process reduce the potential for errors ever reaching the patient.

Once the final ingredient amount is confirmed to be accurate via the electronic scale, the technician will inject the ingredients into the carrier fluid and weigh the bag for a final check. Only upon successful completion of this final step will a medication label be printed on the wireless battery-enabled portable printer, signifying validation of a successful preparation.

Once visually checked by the pharmacist, the preparation of the chemo is complete, and it will be readied for delivery to the pediatric unit. This complex process occurs thousands of times each day in hospitals across the world. Each step of this process has been documented and stored in the computer generating an audit trail of each step. Analytics on productivity, drug consumption, safety issues, and potential inefficiencies in the workflow are now available to pharmacy leadership. This tool not only helps catch and prevent compounding errors but also helps meet compounding guidelines like USP797. The medication now waits to be delivered to the pediatric unit, where it will eventually be infused using a smart infusion pump like the one you saw a few minutes ago.

Emergency Kit and Tray Vending

The restocking process for emergency kits and trays utilizes an automated RFID system. When a tray is restocked, the tray is stored in this smart vending cabinet ensuring appropriate par levels are maintained.

When the pharmacy technician is notified to replace a specific kit in one of the units, such as the ED, the technician's proximity badge grants access to the cabinet and the desired kit is selected. The smart cabinet identifies and records which kit was removed, and the technician defines the deployment location by selecting the delivery location "Emergency Room" from an approved location list. It's quick, accurate, and requires no change in pharmacy workflow.

RoboCourier

These items are now ready for delivery. Instead of staff walking them to the patient areas, an autonomous mobile robot will transport them. The pharmacist logs into the robot with a swipe of their identification badge, opens the doors, and places the medications inside. After closing the doors, the desired destination buttons are pressed on the touch screen and the robot navigates safely and securely to a nurse on a specific nursing unit, a dispensing cabinet, or directly into a patient room. Along the way, the robot rides elevators, opens doors, navigates around obstacles, and updates the status and location of the medication, providing full chain of custody and real-time visibility to nursing and pharmacy staff, allowing them to react quickly to changing conditions.

Medication Packager

In hospitals throughout the United States, it's important to follow the Joint Commission standard to provide medications to nurses in a barcoded, ready to use form and in the smallest unit of measure. Even today, though, medications still come to us in big bottles like this. An automated packer system takes tablets and capsules from bulk bottles and, either in a batch process or on-demand, will repackage each one into a specific unit dose package.

Similarly, many injectable medications come in large multi-dose vials that have more drug than what is needed for one dose. As a simpler alternative than a multi-dose vial, these ready-to-administer, prefilled injectable syringes are available from FDA-regulated manufacturers who prepare drugs using GMP, otherwise known as good manufacturing practices. They are designed to improve patient care and safety by decreasing the number of steps in the traditional model of transferring contents from a vial to a syringe. One of the most cumbersome steps in this process is labeling the syringe that you've used to draw up the medication, a Joint Commission requirement. With unlabeled syringes, the potential for mix-up is big! With syringes like these that are clearly labeled for easy identification, there's absolutely no risk of ever having an unlabeled syringe because we can eliminate that cumbersome step.

In addition to better labeling, simply having a smaller package like these promotes safety because if an error does occur, the potential for harm is less because the dose is less. And before the nurse gives these medications to the patient, scanning this barcode with a bedside barcode scanner attached to the EMR, we can make sure the five rights of medication administration are present: right patient, right drug, right dose, right route, and the right time.

Controlled Substances

Regulatory agencies, such as the Joint Commission, closely monitor and review the processes of how controlled substances like morphine and Vicodin are managed. Controlled substances are some of the most commonly used medications within any hospital. Being that they are controlled, there are many regulatory requirements necessary to remain compliant and in good standing with the Drug Enforcement Administration (DEA). In the old days, if controlled substances were tracked, they were generally tracked by storing them in locked cabinets or cupboards on the nursing unit and two nurses counted all of them, each shift, recording the beginning and ending counts on paper logs. This process took a lot of time, and if a discrepancy was ever detected, it was difficult to determine the reason, making it inaccurate and putting the hospital at risk of diversion of controlled substance medications.

Designed to store, manage, and track all the controlled substances, this system electronically receives replenishment requests from the automated drug storage cabinets out on the nursing units. The system then queues up the necessary items needed to replenish the cabinets before they stock out, debiting the sending cabinet you see here and crediting the receiving cabinets on the nursing units once the medication is delivered. Much like the way FedEx tracks packages, medication barcodes on the packages are read at each step of this process. Once these controlled substance medications reach their destination throughout the hospital, the process is closed. But if, for example, something doesn't arrive as expected, a discrepancy is created along with a full history of the staff involved in the process. With a system like this, discrepancies are detected quickly, investigated fully, and reconciled completely.

Analytics

To manage the inventory of medications and even supplies, it is important that one employ a reporting tool that properly consumes data from multiple disparate sources and combines it into one, simple view of a hospital's unified inventory. It is in this manner that hospitals have a hope of monitoring inventory across the care settings and ultimately across their hospital system in the case of multi-hospital systems.

Temp Monitoring

Wherever medications are stored in a hospital, whether it be the pharmacy or the nursing units, it's a requirement of the Joint Commission and other regulatory agencies that medications be stored within their appropriate temperature range. Because many medications require refrigeration, pharmacy staff must monitor, track, and report on the temperatures of the medication storage refrigerators. Doing this manually twice or three times a day is time-consuming, error prone, and a poor use of highly trained personnel's time, not to mention only two or three times a day. If a refrigerator goes out of range, it could be hours before it's noticed and remedied.

Automated continuous temperature monitoring allows the pharmacy staff to monitor temperatures on any time interval desired and track temperatures electronically. When temperatures go out of range, the system proactively notifies appropriate personnel, reducing the chance of pharmaceutical spoilage. It also allows for corrective actions to be documented, complying with Joint Commission audit requirements.

Business Intelligence dashboards provide a quick overview of the environmental conditions of the refrigerators in the pharmacy and throughout the hospital and save valuable time in day-to-day pharmacy operations, staff productivity, improved regulatory compliance, and ultimately improved quality of care.

Time Synchronization

And the Intelligent Hospital doesn't just monitor the environmental conditions of refrigerators. We also continuously monitor the environmental conditions of the building itself, like temperature, humidity, and differential room pressures. Through a robust network and wireless infrastructure, a dashboard for refrigerators, we can visualize environmental conditions throughout the hospital. And with so many independent devices throughout this hospital, all feeding data into a master record correctly, the Intelligent Hospital has deployed a Wi-Fi-based time synchronization system using Network Time Protocol. This guarantees that all clocks, and NTP-enabled systems like medical devices, are synchronized throughout the entire hospital.

Analytics

In pharmacy, we have a saying that if someone wants to divert medications, they will find a way. Using a Controlled Substance Management System like the one you saw earlier is effective for retrospectively tracking medications from the pharmacy to the automated dispensing cabinets throughout the hospital. But to catch diversion of controlled substances, tracking is simply not enough. No system is perfect, and we must always be careful and proactively look for clues that might indicate diversion is occurring.

The system you see on the monitors now analyzes the usage patterns of all staff who have accessed controlled substances throughout the hospital. It is designed to compare users against each other and help identify telltale patterns of diversion, like excessive vending and excessive wasting. With this system, it's just a matter of time before diversion is detected. Impaired employees are not only a danger to themselves but also a danger to the patients they are actively caring for. And if you don't look for it, you can have a big problem happening and not even know about it until it's too late.

Outpatient Pharmacy

We will now shift your attention to the Outpatient Retail Pharmacy Area behind you. An employee has just arrived to pick up her prescription that was called in earlier today by her doctor. For employees who want to use this pharmacy, a swipe of their ID badge is all that is needed to make a cashless purchase. And not limited to just the pharmacy, the badge also works in the cafeteria, gift shop, and all the vending machines across the campus. If the employee decides to grab a magazine and some gum, the total amount she spends will automatically be deducted from the next paycheck.

Previous Transaction Audit Check

Every retail environment needs to keep close tabs on what is happening at the registers. In this case, if the employee reviews the purchases from a web browser and notices that next week she was overcharged, there's a full audit trail and surveillance of the transaction. A manager searches the system by name and date and brings up video footage and a receipt of the transaction. Here you can see that indeed only one magazine was purchased, but the receipt shows two. The manager corrects the situation and credits the customer for the error. A cashless system with point-of-sale surveillance allows you to manage retail activity with tight security controls in place.

War Room Workflow

As healthcare institutions transition to provide an increasingly integrated and seamless healthcare environment with information being collected, collated, processed, archived, and distributed throughout the healthcare

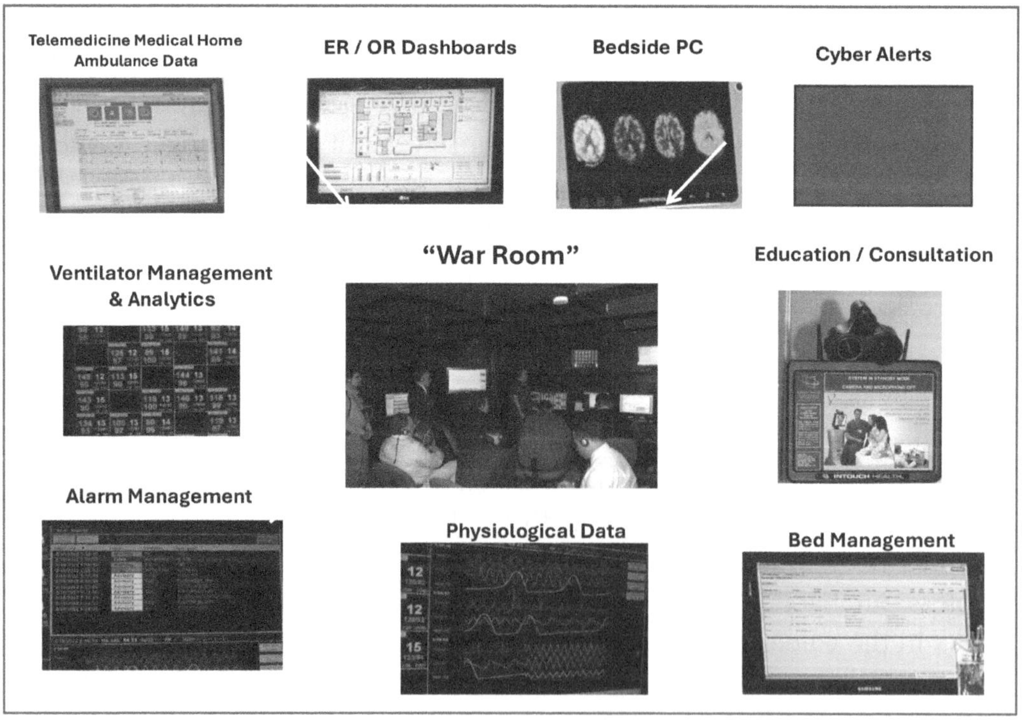

Figure 2.9 Central war room monitoring.

network or system, the ability to centrally view, validate, review, and archive information has become increasingly important. There has been an evolving concern that there should be a location where data coming from multiple places can be collated, filtered, and displayed at the patient, unit, or hospital wide level. With this extensive data flow architecture, monitoring the data from specific locations can be critical. In many cases, the workflows associated with patient care within specific care areas, as well as monitoring institutional workflows and operations, have become difficult increasing the potential risk to patients. In addition, as hospital operations increasingly cross departmental crossing unit boundaries, as in the case of patient data from an ambulance of patient enroute to ED. Consequently, increased consideration for the dynamic review of large volumes including alarms generated from many locations has driven the need for a centralized command center or war room (Figure 2.9) to be a critical component as part of the design of the Intelligent Hospital. The Command Center enables a complete visualization of the events occurring across the institution providing redundancy and oversight that all events are properly managed, and that critical conditions and alerts are not missed. The Intelligent Hospital architecture ideally includes a centralized war room configuration or mission control area to provide the necessary oversight and visibility across the entire healthcare system. Further driving the need for this Mission control function is the evolution of home-based medicine and monitoring. As we move more and more toward telemedicine and home health applications, the Intelligent Hospital will be required to develop strategies and a process where information from these remote locations is monitored and reviewed, and care decisions are made. This room functionally gathers and consolidates information from all areas of the hospital and enables information collated for real-time monitoring, data review, and analysis.

Hospitals typically take one of two directions in managing real-time patient information. One is a distributed model where data is delivered to the point of care, and the other is a centralized data repository, typically manned by experts in the evaluation and interpretation of the incoming and critical information. Many institutions deploy a mixed mode providing a War Room for institution-wide oversight, with unit level oversight at the nursing stations within the specific units, and direct bedside or point of care delivery for interpretation by assigned staff. The specific visibility provided is based on the medical systems, clinical applications, and integration that exist within the specific care areas and can be summarized as follows.

Patient Dashboard

An enterprise-wide real-time census dashboard can visualize the status of the entire hospital. Patient census can be represented by department and include the bed status as a function of utilization and availability. The system can compute aggregate occupancy metrics for the entire enterprise and provide specific information about all the patients including room assignments and actual current location. Leveraging the power of RFID, this dashboard can track the movement and status of every patient, care provider, and asset throughout the hospital.

Physiological Monitoring

Displaying patient physiological monitoring information enables the concept of a clinical information cockpit with access to critical information collected by the bedside monitors and telemetry transmitters. Patient information collected in the ED, ICU, and the Medical/Surgical Unit can be easily monitored and reviewed and subsequently compared to previously collected ECG information to support informed care decision-making. Alarms and alerts can be monitored to ensure staff respond to critical events enhancing patient safety.

As an example, a patient is admitted to the Medical/Surgical Unit for observation as a result of their injuries. Traditionally, patients in a Medical/Surgical Unit only have vital signs spot-checked at 4-hour intervals. This leaves these patients at risk for complications of respiratory failure, hemorrhage, fall risks associated with trauma, and weakness. In the War Room, physiological parameters are continuously monitored. The patient transmitter continuously records her Heart Rate, Respiratory Rate, Oxygenation, end tidal carbon dioxide, and blood pressure. This information is then available as real-time waveforms, vital signs, graphical trends, and alarms. This system also automates the collection of vital signs on the Medical/Surgical Unit with easy integration to the hospital's EMR system. With defensive monitoring, clinicians are alerted when the patient's condition starts to deteriorate, enabling quicker interventions to prevent in-hospital complications leading to better outcomes.

Surgical Information Systems

Driven by scheduling and clinical documentation from today's surgical cases, the visual communication center captures and conveys critical patient data throughout the perioperative department and beyond. The communication center tracks patient location and progress, ensuring schedule adherence. The entire care team is informed about changing patient locations and typical precautions and conditions. The system displays current milestones that reflect a facility's process workflow, and caregivers are notified of updates via mobile messaging.

OR and anesthesia leaders review the days OR schedule via tablet from home or on site. Up-to-the-minute case status for every OR is displayed along with visual cues for important milestones such as scheduled start time, actual start time, and estimated end time. Performance details to see OR downtime, the number of add-on and cancelled cases, and OR turnover time, which indicate how efficiently the OR is running. Proactively managing these changes throughout the day improves patient throughput and workflow.

Role-based analytics provide nurses, surgeons, anesthesiologists, and hospital executives with a clear view of perioperative department performance and actionable intelligence that helps them develop plans to improve outcomes, processes, and efficiency. Advanced drill-down and intuitive data mining features enable the OR control officer to view, explore, and measure perioperative data in just a few clicks.

OR Real-Time Status

Also, within the War Room, direct access to OR surgical information systems are available where live OR video feeds, anesthesia physiologic monitoring, PACS imaging displays, and remote presence robotic consult platforms are all accessible from this critical centralized command center.

In addition, specific functions such as the Sterile Processing Dashboard can visualize Disinfection and Assembly areas providing a clear and complete visibility over the entire surgical instrument workflow process. By leveraging real-time location capabilities and utilizing autoclavable RFID tags, an institution can track and locate both surgical instrument sets and individual instruments throughout the hospital.

Enterprise Visualization

Enterprise-wide operational intelligence solutions can provide a real-time view of the operations of a specific hospital or the entire healthcare system, much like a flight control system provides air traffic controllers with a real-time view of their airspace. Real-time data, collected automatically from IT systems, medical devices, sensors, and multiple RTLS systems throughout the healthcare system, can provide a rich data source for improving healthcare operational efficiency and patient safety.

Optimizing your inventory and asset management enables workflow processes and asset optimized use. Asset Dashboards provide a graphical display of equipment by time in status for each location. Looking across the entire healthcare system, one can identify key factors, such as in certain locations IV Pumps, that spend a large percentage of their time in a "dirty" status, meaning that there might be an opportunity for process improvement. The inventory of IV Pumps in another facility, on the other hand, appears to be well stocked. In fact, they may have more pumps than needed which presents an opportunity to improve pump management and reallocate some of the pumps to one of the other areas. Note that this data was collected automatically, aggregated, and processed across the enterprise and that similar enterprise-wide analytics can be run for patient workflow, environmental monitoring, patient safety, infection control, etc.

The Intelligent Health System, if equipped with smart oxygen regulators monitoring the location and pressure status of all oxygen tanks, can provide alerts and circumvent critical events. Analytic reports illustrate and provide complete transparency of tank status allowing for increased patient safety, optimizing maintenance schedules, and reducing tank rental costs.

The same type of visualization and real-time information can be applied to enterprise environmental monitoring, including temperature monitoring. Not only can you view the status of your temperature-monitored items, but you can also document temperature compliance to regulatory agencies.

Supply Chain Management

Material managers can check and verify that procedure rooms and central storage are stocked with specific inventory at the appropriate par levels or to support a specific procedure. The user can locate and check specific smart cabinets from the department map and verify the inventory contents. Missing or low stock items/products can be replenished and/or ordered to avoid disruptions to patient treatment or the care process.

Medication Management and Administration

Timely and accurate delivery of medications is crucial for patient care as well as patient satisfaction. Tracking medication deliveries is also valuable to the pharmacy and nursing staff to reduce the time spent searching for missing medication deliveries. Maintaining chain-of-custody and providing real-time medication delivery tracking provides pharmacy operations and nursing staff the assurance that the right medication is delivered on time to the right location and to the correct caregiver. The system tracks medication deliveries through multiple means including pneumatic tubes, manual courier deliveries, and mobile robots. The ability to track RFID labeled as well as standard barcode labeled medication deliveries enables the accurate status of the medication at each stage of the medication administration process.

ED Visibility: Patient and Staff Flow

In addition to asset tracking, RTLS solutions offer visibility into patient flow in the patient care areas, such as the ED. The ED is a complex environment with many moving people and resources. Knowing the real-time location and status of clinical staff, patients, and rooms is essential to improving patient care and departmental workflow. Lack of such visibility can slow patient throughput and adversely impact care delivery and outcomes, resulting in extended wait times and unnecessary labor costs. Combining the location and status data allows you to have true visibility into the flow and operation. RTLS enables the rapid location of patients, staff, and available equipment as well as room status indicated by color codes – all on one dashboard display. Customizable alerts can be set to indicate long patient wait times, patient movement between departments and patient calls for assistance, etc., all enabling optimization. Analytic reports enable hospitals to analyze and optimize patient flow. These reports can drive business decisions based on actual data, including patient wait times, patient-staff-equipment interactions, patient and staff historical location, room/bed utilization, and turnaround times, thus improving overall operational efficiency.

Clinical Surveillance Monitoring

A state-of-the-art clinical war room includes a context aware, EMR-enabled, bunker solution consisting of multiple integrated displays. It allows one remote clinician to provide oversight for up to 50 patients utilizing high-fidelity data from all bedside devices. A matrix type of display can provide a quick summary view of primary patient measurements and alarms. A patient centric detail view can be selected providing in-depth analysis including waveforms, trends, analytics, and the ability to configure patient-specific intelligent alarms. An alarm view allows for the display and prioritization of device-generated patient alarm conditions and intelligent alarms generated by the system. Coupled with a two-way video/audio link provides a second witness for documentation and best practice. The systems essentially enable virtual patient rounding which allows for patient assessment and immediate notification and assistance by the bedside caregiver.

Nurse Call

Today's advanced nurse call systems provide functionality designed to increase safety, enhance clinical workflow, and improve patient satisfaction. To achieve desired outcomes, it is critical to capture specific metrics such as the number of requests placed, how care was delivered, and how the response times compared with the team's care targets. Reporting captures virtually every event that occurs within the Nurse Call System. This is measured from call placed to call cleared including all types of patient calls, call reminders, and workflow events like room status for bed flow, staff arrivals, and departures for safety and efficiency. Utilizing an intuitive interface, all nurse call events can be filtered to one central location, thus allowing caregivers to answer, triage, and dispatch the right caregiver to the right patient at the right time from a central location, such as the war room.

A central war room configuration provides a redundant oversight of the many parallel operations occurring across the many patient care areas of the Intelligent Hospital. As hospitals increase their operations to telemedicine and home care, the war room becomes increasingly important to provide remote patient care and decision support and recommendations. As a simple example, observing changes in a patient's physiological parameters or levels of activity will enable care providers to make an early decision to recommend follow-up, ambulance or emergency care, or admission to the hospital. As the acuity of the in-patient environment continues to increase, the use of remote monitoring with become a key functionality to enable early detection and support of critical issues developing.

Section 3

Engineering and Technology Development: Patient-Specific Solutions

One of the fundamental objectives of the Intelligent Hospital is to explore and investigate new and evolving technologies and assess how these technologies can enable enhanced functionality and operations within the patient care environment. The Intelligent Hospital is designed to support a patient-centric model where the treatment is focused on the patient and commonly carried out at the point of care. The creation of this seamless environment has further expanded the opportunities to leverage patient information and data to develop clinical patient-centric and patient-specific solutions. The patient-centric care model has been highlighted in previous sections, describing how treatment, care, and management are delivered at the patient's bedside, remote location, or in the home. Analysis of the aggregated data has further identified possibilities for utilizing data to develop patient-specific solutions. These solutions are enhancing or targeting a patient's unique information to develop or tailor treatment modalities focused on their physiological requirements such as genetics, chemistry, and anatomy.

3D Visualization and Design of Patient-Specific Solutions

Leveraging advanced 3D technologies including imaging, reconstruction, 3D printing, and augmented/extended reality (AR/XR) applications can significantly impact and enhance patient outcomes, treatment efficiency,

DOI: 10.4324/9781003414568-3

and care. These improvements align with the primary objectives of the Intelligent Hospital.

In recent years, new imaging modalities and applications have been adopted within the healthcare environment to provide both static and dynamic 3D reconstructions and visualizations of anatomic features. When coupled with a variety of technologies and applications – e.g., medical 3D printers and/or AR/XR – these capabilities provide enhanced capability in visualization, treatment planning, shielding, radiation therapy (RT) immobilization, and surgery.

These digital technologies can help to define an advanced patient care modality which meets the strategic objectives of the Intelligent Hospital. The goal for employing these tools is to create patient-specific solutions that enable and support multiple treatment modalities which can enhance patient outcomes, safety, and satisfaction. These solutions cross the boundaries of many clinical applications and departments. Transitioning this technology into clinical practice requires institutional support, planning, and strong operational synergies between Radiology, Biomedical Engineering, and clinical users, such as Surgery, Dentistry, Radiation Oncology, and others.

Since digital reconstructions are based on specific anatomy derived directly from a patient's imaging, an anatomically accurate representation enables a variety of tailored solutions to be developed. Some of these advances, as summarized below, have enabled physicians and clinical staff to develop unique solutions based on individual patients' anatomy and treatment needs. These solutions can enhance treatment and improve patient outcomes.

- Patient-Specific Anatomically Accurate Models
- Procedure-Specific Surgical Tools and Cutting Guides
- Dental Guides and Tools
- Patient-Specific Immobilization Devices
- Custom Radiation Shields
- Direct Targeting Devices
- AR/XR Visualization

The following sections highlight a broad set of applications and use cases and describe the considerations and planning of these technologies as part of the development of the Intelligent Hospital.

Patient Custom Radiation Therapy Immobilization Devices

External Beam Radiation Therapy (EBRT) was one of the earliest drivers defining the need for an expanded role of Biomedical Engineering within the patient treatment space and is still an important area of focus today. EBRT delivers high-energy electron, photon, or proton beams to the region of a patient's tumor(s). The dosages, exposure, and beam orientations are defined by radiation oncologists and medical physicists by creating a unique treatment plan with an associated orientation and position. The treatment plan dynamically positions the linear accelerator and targets the high energy radiation delivery to the designated points of interest (tumors).

Fundamental to the RT process is to ensure the patient position is uniquely defined for the treatment plan geometry. Custom positioning and immobilization devices are designed to ensure a patient is properly orientated as defined in the treatment plan and that the position is precisely maintained over the course of multiple therapy sessions. The patient is immobilized during the radiation delivery and precise radiation beam targeting is coupled with radiation shielding to limit the radiation exposure to non-tumor-related healthy tissue.

Historically, these immobilization devices were made by hand and required a patient to be in the room to set up and simulate treatment positioning prior to treatment planning. With modern digital tools, it is now possible to prepare immobilization devices from diagnostic imaging and manufacture immobilization devices using computer-aided design (CAD) and computer numerical controlled (CNC) machining. An example of digital immobilization device production is shown in Figure 3.1. These CNC milled devices can be prepared without the patient coming in for a special appointment and they can provide added comfort and tailored features which were not previously possible.

Custom Patient Shielding

Shielding blocks are mounted in the linear accelerator which then limit or minimize exposure of non-tumor-related healthy tissue. Shielding materials include but are not limited to lead, lead alloys (such as Cerrobend), copper, brass, tungsten, and other heavy metal alloys. As illustrated in Figure 3.2, custom radiation shielding blocks can be manufactured – in this case using Cerrobend material –directly designed and manufactured from patient

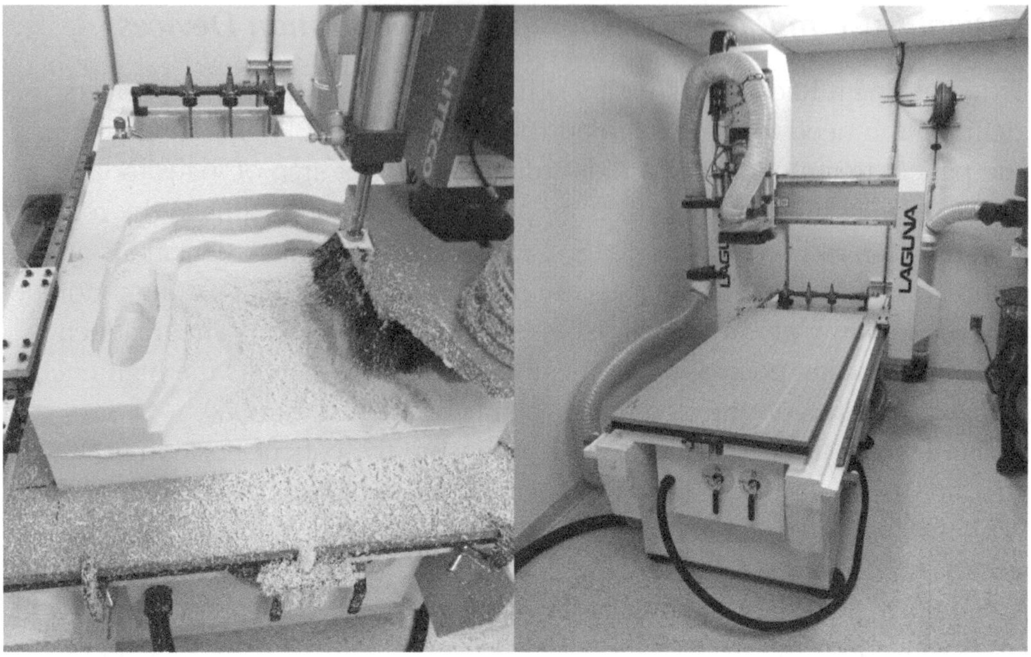

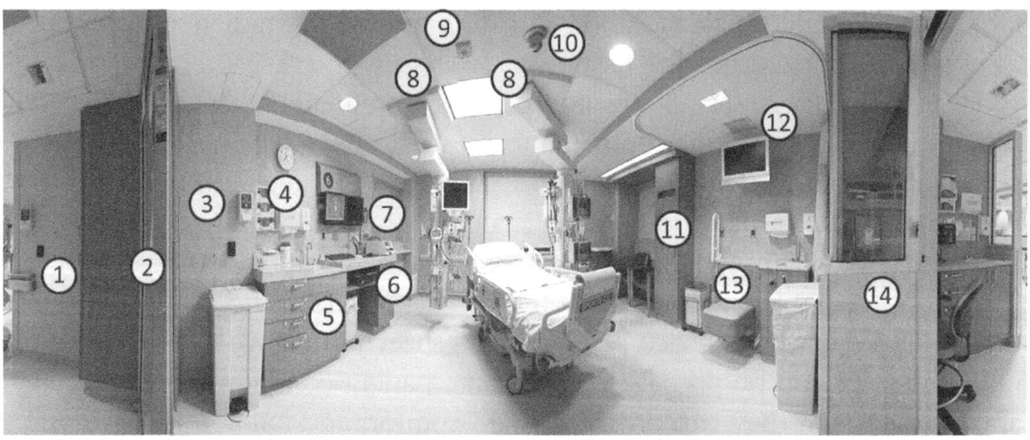

1. Nurse server & secure ID card access reader
2. E-glass slide & breakaway doors
3. Inside opening of nurse server
4. Wireless Time Synchronized clock
5. Storage cabinets
6. Bedside computer with double monitor
7. Barcode reader and lab label printer
8. Twin mobile articulating arms (BOOMS)
9. Wireless IR transmitter
10. Web cam
11. Closet and DVD player
12. Flat screen TV on articulating arm
13. Toilet
14. Nursing work area
15. Antimicrobial surfaces

Figure 3.1 Custom patient-specific immobilization device (left) being milled on a CNC router (both images). (Critical Care Unit Layout Memorial Sloan Kettering Cancer Center.)

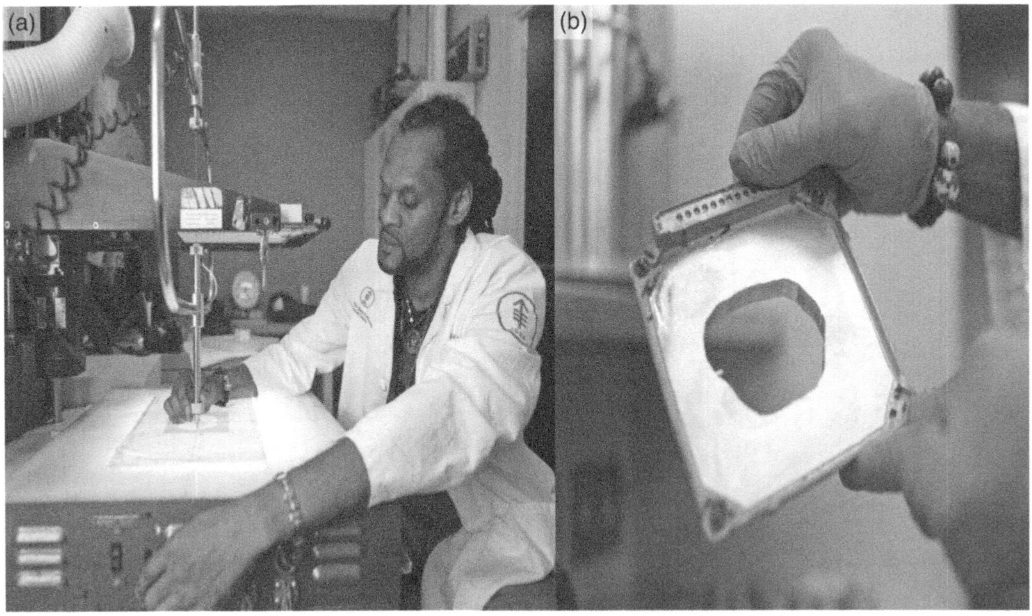

Figure 3.2 Custom patient shielding. (a) Technician tracing treatment plan from patient imaging with a tool that replicates the electron beam path. (b) Finished shield made of Cerrobend, lead alloy.

imaging with the shielded areas defined to meet the objectives of the treatment plan.

Mechanical Engineering – Instrumentation Shop

The requirements for creating patient-specific devices have resulted in the need for Biomedical Engineering functions beyond the traditional clinical engineering role assumed within the hospital environment. This has established the demand for Mechanical Engineering roles as part of Biomedical Engineering and patient care. Along with digital imaging and production, a robust suite of tools can be incorporated into point-of-care design and manufacturing. There is a wide range of computer aided design/manufacture (CAD/CAM) software suites and a significant role for skilled design engineers and technicians. In addition to digital production, tools can also include more traditional subtractive manufacturing – e.g., machining such as milling, drilling, and turning – and other hand fabrication such as forming, welding, and others.

As illustrated in Figure 3.3, a mechanical engineering capability may include a facility capable of machining and fabrication for both production and prototype applications. The shop operations support CNC and manual

Figure 3.3 Mechanical instrumentation shop.

mills, lathes, saws, forming tools, sandblasting, etc. to enable manufacturing for metals, plastics, wood, and composites.

Additive Manufacturing – 3D Printing

Additive manufacturing (3D printing) has brought a new dimension to Biomedical Engineering in the Intelligent Hospital. Unlike traditional mechanical fabrication techniques based on the removal of material (CNC and milling machines) to create specific objects, additive manufacturing (3D printing) is defined as the construction of three-dimensional objects from a CAD model or other digital model. 3D printing can utilize a variety of processes in which material is deposited, joined, or solidified under computer control with material (plastics/metals/biologics in liquid, solid, or powder forms) being added and fused together and objects constructed layer by layer. 3D printing encompasses a wide range of technologies that have historically been emphasized for their application in rapid prototyping. However, there has also been a steady increase in the adoption of 3D printing for creating end use products. In the hospital environment, it has become widely adopted both for its traditional prototyping capabilities and for its ability to produce a range of clinical, diagnostic, and educational tools. 3D printing typically has limited mass-production throughput; however, it represents a very viable option for low production numbers, specifically targeting custom solutions, and is extremely flexible when compared to traditional fabrication technologies and technologies. It can allow for the creation of a wide range of products and devices without the overhead and lead time required for other methods.

Rapid Prototyping

Within the Intelligent Hospital, 3D printing has become widely adopted both for its traditional rapid prototyping capabilities and for its ability to produce a range of clinical, diagnostic, and educational tools. 3D printing has enabled flexible designs which are more complex than are possible from traditional manufacturing or milling processes enabling the rapid prototyping of complex shapes including hollow parts or parts with internal truss structures for reduced weight objects.

General Hospital Support Services

This flexible capability has significant advantages and impact on applications which address the operational needs of the hospital but are not necessarily medical devices or used directly in clinical treatment. Some items which are used in the patient care environment do not directly influence patient care; however, they can directly impact operations and/or operational efficiency. While the clinical and R&D applications are currently more suited to large research hospitals, these support services are finding use even in smaller hospitals throughout the world.

Having 3D printing capabilities in a hospital allows flexibility both under normal operations and in the case of emergencies or supply chain issues. We have found many applications to improve workflows and patient care by filling in the gaps where commercial solutions are not available. For example, custom holders have been printed to keep wires organized at patients' beds or unique ultrasound probe holders printed to place them in a more accessible location for our room layouts. Custom mounting brackets have been printed to enable RFID identification tags to be mounted in inconspicuous locations and improve operational capabilities or prevent damage. These applications are typically too time-consuming or expensive for hospitals to fabricate and are commonly outsourced; however, they become relatively simple to implement once 3D printing is available. In times of crisis, 3D printing provides the ability to respond to unforeseen needs with flexibility and speed. For example, during supply chain shortages seen in the early stages of the COVID pandemic, our institution used 3D printing to make face shields, nasopharyngeal test swabs, viral filter adapters, and respiratory fittings as shown in Figure 3.4 [1].

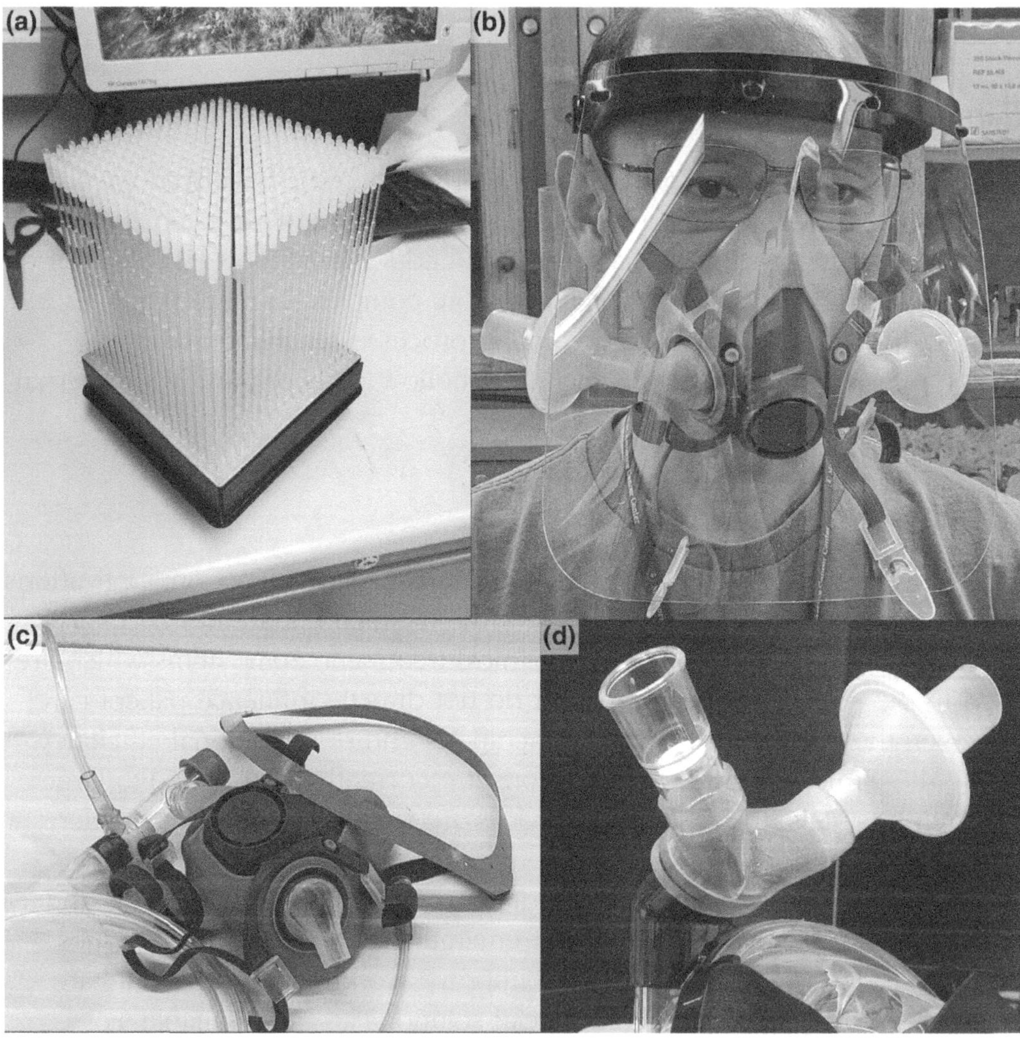

Figure 3.4 (a) Tray of 3D-printed nasopharyngeal swabs used for COVID testing. (b) In-house face shield (3D-printed headband and stamped plastic shield) and viral filter adapter used on half-face respirator (allowing use of anesthesia viral filters when the hospital ran out of proprietary filters). (c) View of respirator and viral filter adapter. (d) Example of 3D-printed respiratory fitting.

Rapid Prototyping

The use of 3D printing provides an excellent tool for design and development applications and enables the production of prototypes for testing and evaluation. Consider the design of an advanced surgical instrument as illustrated in Figure 3.5. This device is a uterine manipulator designed for use in gynecological surgery. Rapid prototyping and point-of-care manufacturing can move a device like this from the design and development phase to the fabrication phase much more quickly. In this case, the transition from the design phase to a production version ready for clinical trials was completed in months as compared to years required for similar devices using more traditional tools and off-site development. Designs are typically based on a collaborative effort between healthcare providers (HCPs) and mechanical engineers using CAD software to design the instrument. The initial design is typically produced in plastic or resin on a 3D printer and enables the HCP to dynamically interact with the instrument. In the uterine manipulator example, multiple design changes were introduced over 5+ iterations. The iterations were quickly produced using 3D printing technology enabling the surgeon to rapidly test and re-evaluate the devices for each iteration and augment the design as necessary. Once the surgeon was satisfied and confident of the design,

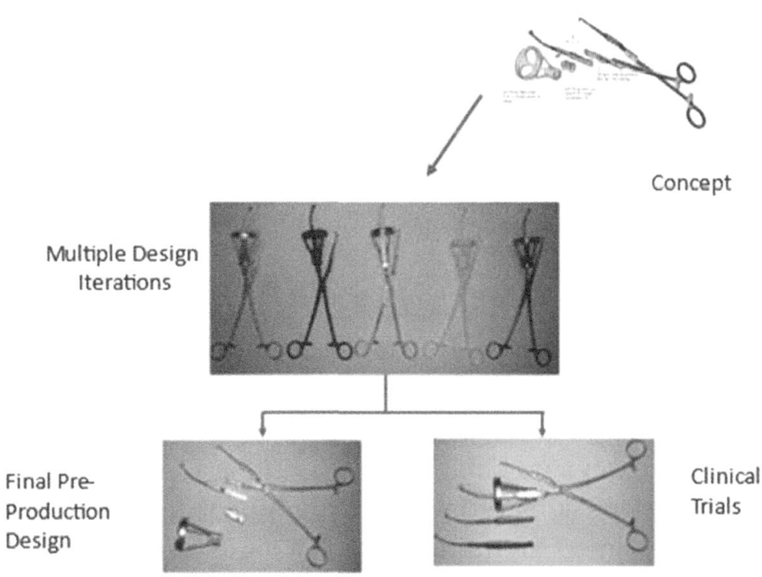

Figure 3.5 Uterine manipulator for gynecological surgeries – showing process from concept to production.

a pre-production instrument was fabricated in metal for further testing and evaluation and the start of the clinical trials process.

3D Printer Systems and Technology

The following provides an overview of some of the common types of 3D printers which are currently commercially available. There is a significant amount of literature, including books, journal articles, and manufacturer specifications, available to provide detailed reference and review of these systems [2]. The choice of printer technology and manufacturer is a business decision based on planned function or projected applications.

3D Printer Technologies:

- Stereolithography (SLA)
- Selective Laser Sintering (SLS)
- Fused Deposition Modeling (FDM)
- Digital Light Process (DLP)
- Multi Jet Fusion (MJF)
- PolyJet
- Direct Metal Laser Sintering (DMLS)
- Electron Beam Melting (EBM)

Printer Functionality and Capabilities

Each of the outlined printers can be used for clinical, research, or operational applications with the choice dependent upon the requirements of the specific project. As a rough framework for thinking of where these technologies fit in, FDM printers use extruded thermoplastic filaments to construct parts which are well suited for lower-cost, end use, or functional applications where mechanical properties such as durability, impact strength, abrasion, and chemical resistance are critical. SLA printers employ an ultraviolet laser to cure a photopolymer resin and can produce parts with a high level of accuracy and detail. PolyJet printing works similarly to 2D inkjet printing by depositing droplets of photopolymer resin to construct each layer. These printers are used for producing high-quality, realistic surgical and anatomical models, molds for patient-specific clinical modeling, as well as functional parts which require unique material properties such as flexibility, transparency, or specific textures (Table 3.1).

Table 3.1 Printer Capabilities and Limitations: Additional Details and Overview of Some Common Types of 3D Printers, Highlighting the Advantages and Limitations of These Systems

Process	*Material*	*Method*	*Advantages*	*Limitations*
Fused deposition modeling (FDM)	Thermoplastic filament	Heated extrusion	Ease of use, functional prints, inexpensive	Variable durability of print, build orientation affects the end part, long print times, limited spatial resolution, and limited surface finish quality
Stereolithography (SLA)	Photopolymer liquid resin	Scanning beam UV curing	Superior accuracy, superior resolution, excellent surface finish, inexpensive, a wide range of material properties	Significant post-processing, support placement considerations, secondary curing, potential for warping
Selective laser sintering (SLS)	Powdered materials	Scanning laser	Strong prints, a wide range of materials including metals	Safety/environmental concerns
PolyJet	Photopolymers	Inkjet printhead UV curing	Mixture of multiple materials, a wide range of material properties	Expensive, significant post-processing
Binder jetting	Powdered materials	Binding agent fusing	Fast, can print physically complex parts, color mixing, and large parts	Environmental concerns, significant post-processing, fragile prints

Establishing 3D Printing Clinical Usage and Standard of Care

The vision of how 3D printing will evolve as part of the Intelligent Hospital must take into account a broad set of requirements and ongoing planning. It is essential to ensure an operational pathway as well as an evolutionary pathway to integrate new and evolving technology and applications as they arise. Today, we already see parallel pathways and applications evolving from advanced 3D imaging, including AR/XR, as well as many new use cases and applications in 3D printing. To achieve these objectives, the following considerations must be clearly defined and planned:

- Facilities Requirements
- Printing Operations (Centralized vs Distributed)
- Accuracy
- Quality Assurance (QA)
- Infection Control
- Biocompatibility
- Regulatory Issues and Requirements

Facility Considerations

One of the often-overlooked areas of consideration are the facility requirements for a 3D printing operation. In addition to space and power requirements, careful consideration of other environmental factors is important as well. For instance, it is increasingly becoming apparent that fumes released during printing may be harmful to those working in the area. Setting up proper enclosures and/or ventilation is critical to provide safe environmental conditions for operators and nearby workers and can be justification for a centralized operation in order to consolidate the number of environmentally controlled locations. For some materials in their raw state (filament, powder, liquid, etc.), there are additional concerns including skin irritation, flammability, and/or other systemic toxicity. In the case of powdered materials, proper handling and containment during transfer and cleaning are vitally important as they can easily become airborne and present inhalation and flammability risks. For reactive metals such as titanium and aluminum, these powders can also potentially be explosive. The requirements for each printer type and material can vary, but a careful plan must be put in place to ensure safe storage, handling, and operation. All waste materials must also be disposed of appropriately. As an example, a functional 3D printer facility is illustrated in Figure 3.6 that includes dedicated vented enclosures and houses ten printers along with supporting devices. This creates an environmentally safe and functional work area for supporting staff.

Centralized 3D Printing Laboratory and Operations

Within the Intelligent Hospital, the evolving role of Biomedical Engineering directly supports many clinical patient-care operations, including the development of patient-specific treatment solutions, clinical simulations, and 3D anatomic visualization. These applications require diverse engineering capabilities including electrical and mechanical engineering, computer science, information technology, materials science, software development, and analytics. The specific organizational structure, capabilities, and skills necessary to develop and maintain these evolving technologies are detailed and discussed in Section 4 focusing on re-engineering Biomedical Engineering.

Figure 3.6 3D printing facility.

The decision whether to have a single centralized 3D printing operation and which group or department in the hospital manages and operates the laboratory can vary between institutions and can focus on the specific applications of interest. There is no clear consensus on the best organization for 3D printing operations. A recent survey study of hospital 3D printing sites – primarily represented by large, North American, university-affiliated institutions – found that just over half of respondents reported having a single print lab site versus multiple locations and, of those, 67% identified Radiology as the managing group [3]. In our institution, which has adopted a centralized 3D printing lab, the laboratory is operated and managed by Biomedical Engineering. The primary decision was chosen due to technical resources within the Biomedical Engineering, coupled with the existing mechanical/machine shop housing a variety of milling and fabrication systems and extensive capabilities in 3D CAD design. There were also strong preexisting clinical relationships between BME and Radiology, Radiation Oncology, Surgery, and other key users of this technology.

The development or implementation of the Centralized model has several key advantages. In particular, centralizing allows effective use of financial resources, enabling the purchase of a smaller number of high-quality, centrally managed printers rather than a larger number of lower-cost printers distributed among different areas. A centralized 3D lab can take advantage of economies of scale and a specialized team with more experience to deliver advanced products. While a centralized 3D printing service may reside within varying individual or collaborative departments at various hospitals, it is important that there be a core group responsible for the sustained operations, upkeep, and knowledge required. Input from many areas, including engineers, technicians, surgeons, radiologists, nurses, and others, may be required for any given application and the central 3D printing lab can coordinate clinical workflows and communication between the needed departments as applications arise and are developed.

Accuracy and Quality Assurance

Accuracy in clinical 3D printing commonly refers to the dimensional accuracy of a physical 3D print matching the digital 3D model or patient anatomy within specified limits. Alternatively, it may refer to other characteristics which must be controlled: radiopacity, biomechanical properties, chemical composition, etc. These specifications will depend on the requirements of the use case and have inherent limitations based on the specifications of the printing methodology chosen and the source data. While the same principles apply regardless of what parameters are being controlled, we will discuss dimensional accuracy here for illustration.

The dimensional accuracy requirement of clinical 3D printing depends largely upon the medical indication of the 3D model being used. For instance, applications such as training and education might have less stringent tolerances than clinical uses such as implants and surgical cutting guides.

Error in the 3D print can be introduced due to human decision factors, information translation from one step to another in the process, and mechanical error with the print. To ensure deliverable quality, there are several quality assurance (QA) and quality control (QC) steps that can be taken to validate that the actual output conforms to the planned output.

The most thorough method is to carry out QC checks on each print; this involves measuring the final printed model/part and comparing it to the initial data source. The measurement can be a complete scan (using, e.g., a 3D optical scanner) of the printed part or can entail measurement of critical features of the part using any number of manual or digital tools (calipers, rulers, micrometers, lasers, etc.). Various metrology and scanning technologies can be used to identify areas of discrepancy between the print and the model; however, it is important to note that the scanning/measuring system itself may need to be maintained to ensure QA. Another approach is (in parallel or on its own) systematic QA and involves stepwise checks on the work process. For instance, a 3D print created from an optical scanning of a patient could be validated stepwise by

a. Taking a few manual measurements on the patient if there are sufficient landmarks available.
b. Checking the accuracy of the 3D scanner by scanning a known reference object.
c. 3D printing of an easily measurable object (e.g., a cube with round hole features) along with the final 3D print to ensure the printer is working properly.

These types of measurements assess the performance of the process and help ensure that the final print is correct.

Systematic QA and part-oriented QC should be part of any robust 3D printing operation since clinical decisions/outcomes will often be based on these models/parts. Allowable error must be determined based on critical areas of the process and a QA/QC program will ensure that produced parts fulfill these requirements.

Sterilization, Infection Control, and Biocompatibility

Some 3D prints are intended to come into physical contact with patients and/or be used in a sterile environment. The safety and sterility of the print must be verified before use in these cases. Again, knowing the application and how the print is to be used will define the requirements. It must be determined whether a part needs to be (a) disinfected, (b) sanitized, (c) sterilized, and/or (d) biocompatible (and for which clinical indication). The first step in creating clinical prints which need to meet any of the above requirements is to use a printer and material specifically designed to be used for the purpose. Several printer manufacturers make commercial printers and materials that are designed and intended to be used clinically, but there are widely varying compliance types. It is important to note, for instance, that simply being sterilizable does not mean that a material is biocompatible and does not automatically allow it to be used in a clinical operation. Users must adhere to the specific limitations that the manufacturer recommends for clinical utilization.

In addition to sourcing the appropriate printer and materials for sterilizable prints, it is critical to follow the appropriate Instructions for Use (IFU) which detail manufacturing and processing steps for the material and for the subsequent sterilization technology. QA/QC procedures need to be established to ensure that all of the steps are followed appropriately and that the device is safe for patient use. In addition, even when a material is indicated as biocompatible for the specified application type and all IFUs are followed, the final shape and features of the print may require testing to ensure sterile processing. For example, a part with complex features or small lumens may need to be tested to make sure that the cleaning and sterilization process is sufficient.

Biocompatibility can play a significant role in clinical 3D printing and, as with sterilization, the first thing a facility needs to do is to establish the patient safety protocols for the intended use and abide by all regulatory requirements. There are various standards that a material can meet with regard to cytotoxicity, sensitivity, irritation, genotoxicity, hemocompatibility, carcinogenesis, and so on [4]. Material selection is largely a factor of application and availability of materials for the printer type and model being used.

Regulatory Requirements

It should be noted that devices to be used on patients and/or in a clinical setting must comply with all appropriate regulatory controls; this may include internal groups such as Internal Review Boards (IRB), legal counsel and clinical leadership, or external agencies such as the Food and Drug Administration (FDA) and the Joint Commission. While point-of-care manufactured medical devices may or may not fall under FDA regulation [5], it is advisable to adhere to FDA guidelines and good manufacturing practices [6] when producing clinical 3D prints. It is strongly recommended to create an auditable workflow with tracking of materials and processes along with QA procedures discussed in the previous sections.

The scope and nature of the regulations, like many other areas in medical printing, are largely defined by the intended use of the print itself. While the responsibility for medical use ultimately falls on clinicians, it is the safety apparatus of their institutions and of regulatory agencies which provides the highest levels of protection for the clinical use of 3D prints. As in all areas of medical 3D printing, it is critical to have an in-depth understanding of each step in the process from conceptual model to the final print and ultimately its clinical use.

Workflow

In order to establish a 3D printing program, it is critical to create a well-defined workflow that effectively produces a standardized and consistent product which can be applied in the delivery of patient care. For example, medical 3D printing has enabled the creation of anatomically accurate physical replicas or models as illustrated in Figure 3.7. These models allow visualization of anatomic structures and can provide insight into patient anatomy and pathology. Printer technology enables models to be fabricated using various materials, colors, and even textures to highlight specific anatomic features, such as veins, arteries, airways, and tumors.

In order to develop 3D-printed patient-specific anatomical models, digitization of a patient's real-world anatomical structures must first take place. This leverages standard 3D medical scanning techniques such as MRI, X-ray, CT, or 3D ultrasound to produce a volumetric image of the anatomy. The image components are labeled, via a process called segmentation, to isolate structures of interest and translate the digital information into an STL

Figure 3.7 3D-printed anatomical model of lung and bronchus.

file which is then interpreted by the printer software to create instructions/tool paths for the printer. The 3D printer then prints an accurate full-sized or scaled version of the anatomic structures.

These anatomically accurate models can add several important capabilities as part of the patient care process. The most clinically focused use of these models is as part of the surgical planning process. In the case of cancer surgeries, removal of the tumor mass is the primary objective. However, frequently the tumor position is interleaved with critical portions of the patient anatomy which represents a challenge within the surgical process. These models can enhance the visualization and planning of the procedure by allowing HCPs to better understand the physical arrangement of anatomical structures. As illustrated in Figure 3.8, the 3D-printed skull model highlights the complexity of understanding tumor position and determining the safest pathway for extraction.

The printing of patient-specific anatomical models for surgical planning is an increasingly common application which requires multiple groups to work together [7]. Figure 3.9 shows a graphic representation of this workflow in our hospital; a request is initiated by the Surgical department and goes first to a surgical review with Surgery, Radiology, and, optionally, BME (if engineering/design work is needed). Next, Radiology determines whether existing patient imaging is appropriate or, if needed, new imaging is then completed. Segmentation is then typically done by Radiology; however, it can be done in part or entirely by BME. If additional design work is necessary, BME will do this and then complete and deliver the 3D-printed model(s). Finally, the model(s) must be reviewed and approved by the radiologist and the surgeon.

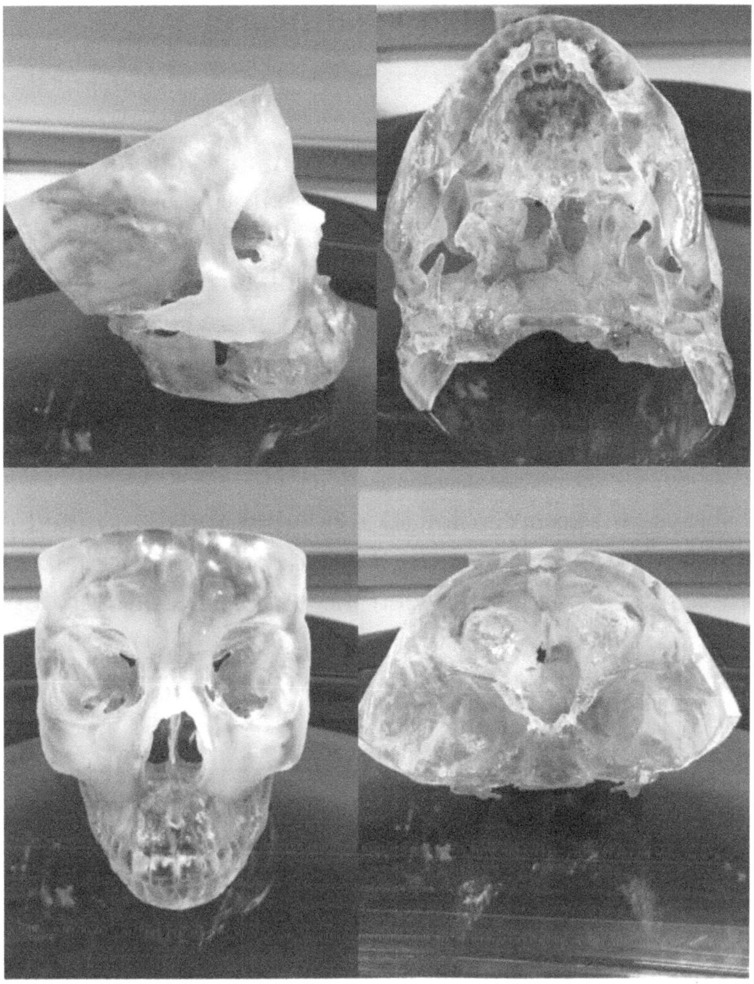

Figure 3.8 3D-printed anatomical model of skull and tumor.

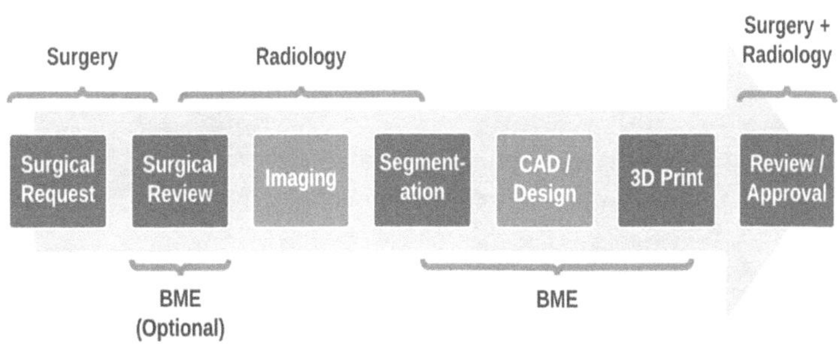

Figure 3.9 Example workflow graph for presurgical anatomical model. Tasks to produce 3D-printed anatomical models are shown left to right, with the responsible group indicated.

3D Printing Applications in Healthcare

Anatomical Models

As described above, one of the most common applications of 3D printing in hospitals are anatomical models used for presurgical planning and in some cases as an intraoperative reference. Research has shown that the use of 3D anatomic models for surgical planning can enhance tumor localization [8], decrease operating room time and cost [9–11], increase surgeons' confidence [12], and improve patient outcomes [13].

Example: Pelvis Osteosarcoma 3D Model

Figure 3.10 shows an example of a 3D rendering and 3D-printed pelvic model with osteosarcoma. Here, the 3D model can help visualize the complex physical relationship between the tumor and the surrounding bone, nerves, arteries, and veins. This type of model can help to plan a surgical approach, reduce surgical time, and avoid complications.

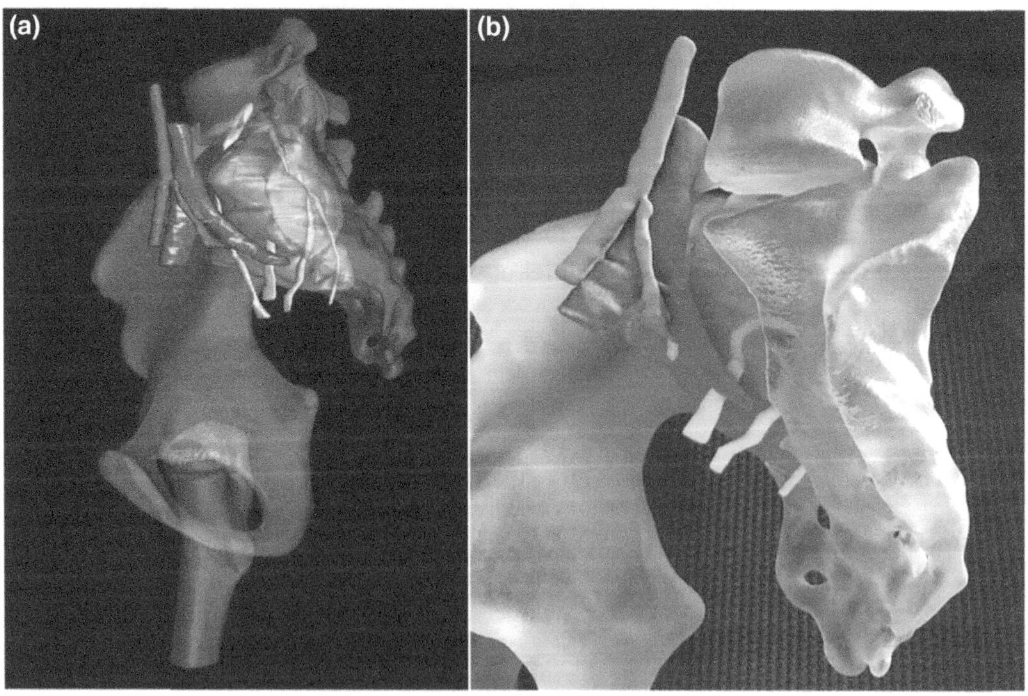

Figure 3.10 3D-printed pelvic osteosarcoma model. (a) 3D rendering. (b) Printed model.

Example: Chest Sarcoma

Another example is a request for a surgical review/planning model of an upper chest sarcoma (Figure 3.11) defined with a full translucent rib cage, trachea and arteries, and tumor.

Surgical Guides

Dental surgical guides are a prime example of clinically used guides in medicine. Patient-specific anatomy is segmented and modeled, and precise clinical guides are created to facilitate reconstructive procedures. Dental

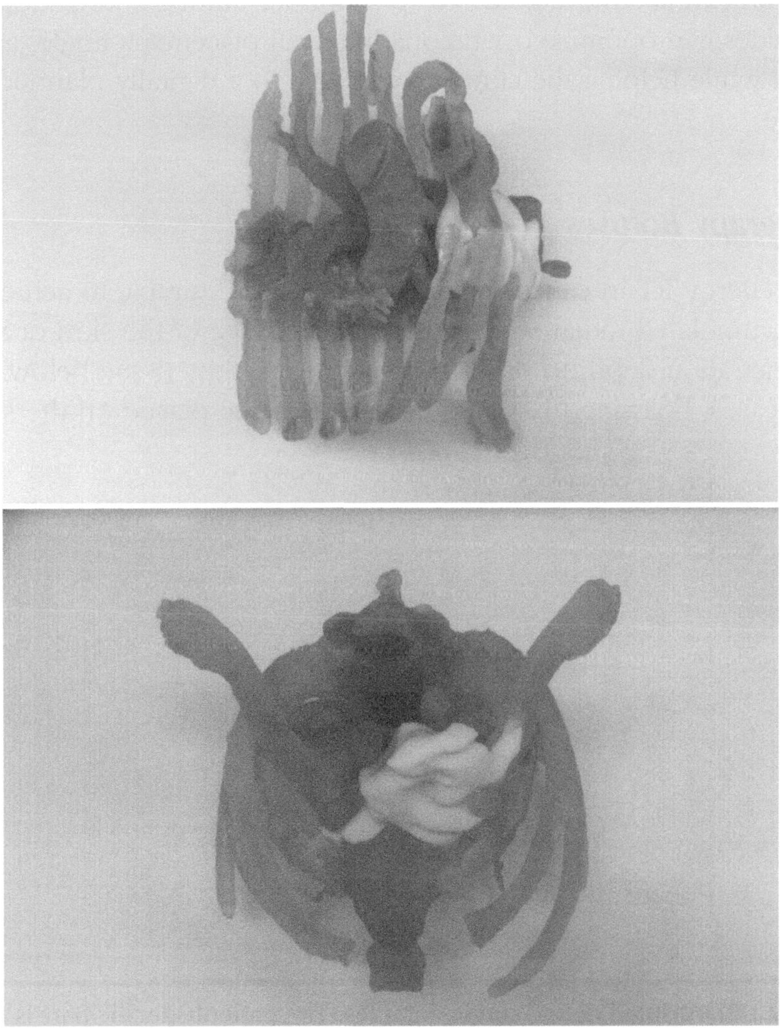

Figure 3.11 3D-printed upper chest sarcoma model.

procedures are particularly well suited for this type of modeling as the teeth provide an accurate and reproducible set of landmarks to ensure proper positioning. Figure 3.12 shows an example of a surgical guide used to properly align drilling locations and depths for dental implants. Here, a biocompatible resin is used and the guide snuggly fits to existing teeth for proper positioning. Press-fit metal bushings are inserted into the 3D-printed model to guide the drill bit.

Typically, bone or teeth are required for use of a surgical guide; however, other landmarks could potentially be used. Depending on the specific application, 3D modeling software is used to effectively create the appropriate model for the clinical application in question. Once completed, the surgeon can use these guides to position for cutting and/or drilling. These guides can optimize operations like drill placement, angle, and depth, all while helping the clinician conform to a digitally planned surgical procedure.

Radiotherapy Boluses

For high energy RT in cancer patients, it is often desirable to achieve higher accumulated radiation dose near the surface of the skin or with a small penetration depth to prevent damaging healthy tissue below the skin. In these cases, bolusing (tissue-equivalent material placed on the skin)

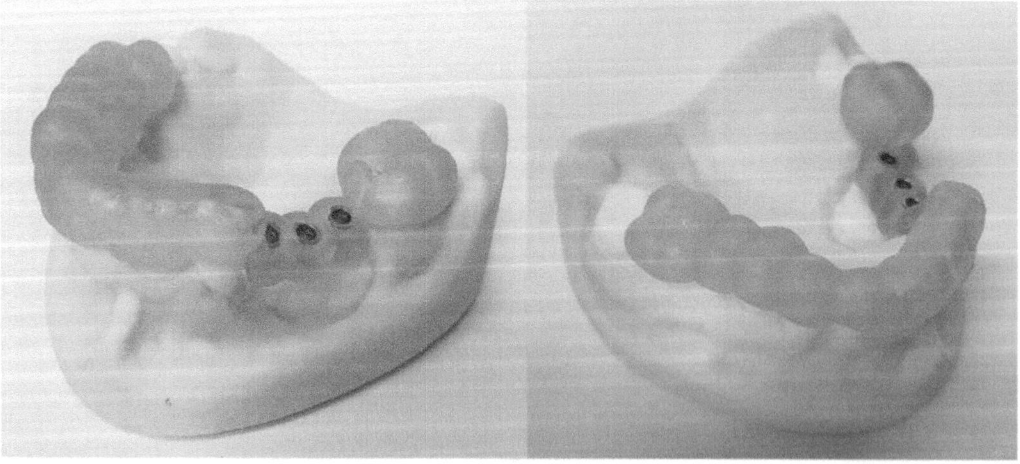

Figure 3.12 3D-printed dental surgical guides. The patient-specific part is the guide which fits over existing teeth, metal bushings press-fit into the guide. Guides shown on the 3D model are based on the accurate modeling of the patient anatomy.

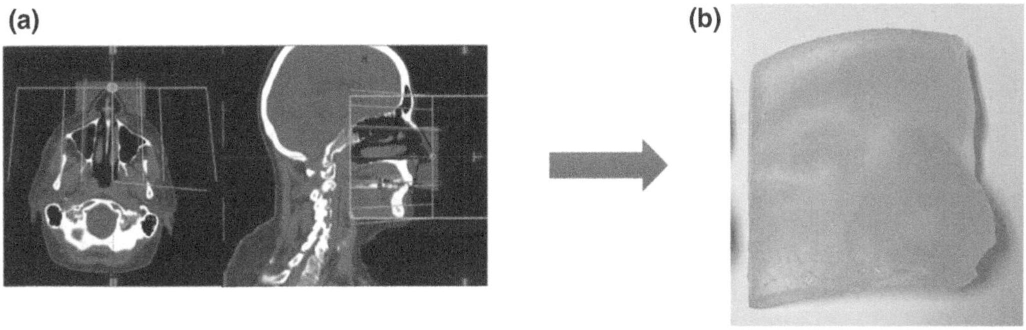

Figure 3.13 (a) CT scan of a nasal cancer patient used for planning and creation of 3D bolus. (b) 3D-printed bolus using flexible material which is approximate tissue equivalent dosimetrically.

is used to achieve the desired dose. Standard sheets of bolus material do not conform well to complex anatomical surfaces and generally leave air gaps which are filled with wet gauze. 3D printing these boluses with an appropriate material allows patient scans (optical, CT, MR, etc.) to be used and provides for an excellent fit with reduced/eliminated air gaps. Figure 3.13 shows an example of a nasal cancer case where a 3D-printed bolus was used to provide proper dosing at the disease site while reducing the dose delivered to deeper tissues (in this case brain). The bolus is printed using a flexible biocompatible (skin contact) resin using an SLA printer.

Brachytherapy Applicators

For rectal and vaginal/cervical cancers, brachytherapy is an important tool to treat disease with local high dose radiation. There are standard commercially available applicators that allow for multiple treatment modalities but do not offer patient-specific adjustment. 3D printing these applicators allows us to tailor the applicator size and shape (to account for anatomical differences or abnormal physiology) as well as needle trajectory (for optimal tumor coverage). Figure 3.14 shows an example of a 3D-printed vaginal brachytherapy applicator. Here, custom configured lumens provide proper positioning of the radiation source for treatment and the applicator is designed to fit a specific patient's anatomy. Both the main applicator and the collets and set screw are 3D-printed from biocompatible (mucosal and breached surface contact <30 days) resin using an SLA printer.

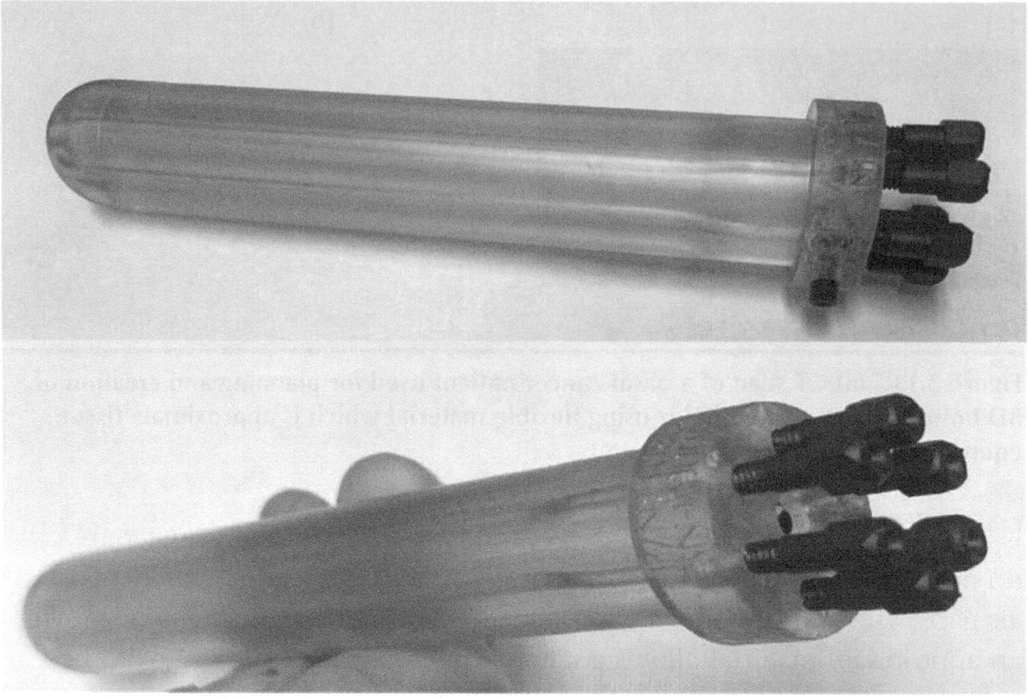

Figure 3.14 3D-printed vaginal brachytherapy applicator.

Emergency Clinical Cases

These are clinical applications which are carried out on an as-needed basis. Typically, these are unforeseen/unique situations which are unplanned and typically do not get repeated (frequently) and consequently are not supported through any standardized workflow process. This can also include educational tools used to train healthcare providers or modeling tools used for patient representation. Examples of non-standard applications can range from a rare infant cardiac defect [14] to aiding in the separation of conjoined twins [15]. While many of these prints are similar to anatomical models, tools, and guides in standard workflows, there are typically alternate requirements, uses, or post-processing needed to complete them.

Example: Oral Prosthesis with Trismus

Patients with oral cancers that have been surgically removed often have a resulting void in the oral cavity. Typically, a prosthetic is specifically created for them by first using a molding material which conforms to the defect and then using that model to create a patient-specific prosthetic that accurately fills the void. Patients who develop trismus can have limited access to the oral cavity and, as a result, some molds may not be accurately formed (or access is virtually impossible). Attempting a mold in these cases can result in a suboptimal prosthesis that doesn't fit and can present a dangerous choking hazard. However, not having the prosthesis has significant clinical and social impacts for the patient as well; speech and swallowing can suffer and result in a host of negative outcomes like social isolation and unhealthy weight loss. In these rare cases, a 3D print of the oral cavity of the patient suffering from this condition can be used. First, we obtain a CT of the patient post-surgically and segment the relevant anatomy into a 3D model, as shown in Figure 3.15. We then 3D-print the model and use it to accurately mold and create a standard prosthesis. The anatomically accurate prosthesis can then be applied to the patient to close the void.

Mandibular Model for Pre-Bending Fixation Plate

Another example of a non-traditional application is employing a 3D model to pre-bend surgical fixation plates. These plates are used to rigidly connect adjoining bone segments with screws. In a specific case at our institution, a fracture in the mandible occurred in weakened bone as a result of surgery and subsequent radiation treatment. The fracture was identified both by the patient and confirmed through CT imaging. A corrective procedure needed to be performed using metal fixation plates which are bent to conform to the specific bone pieces in question and screwed into the mandible on

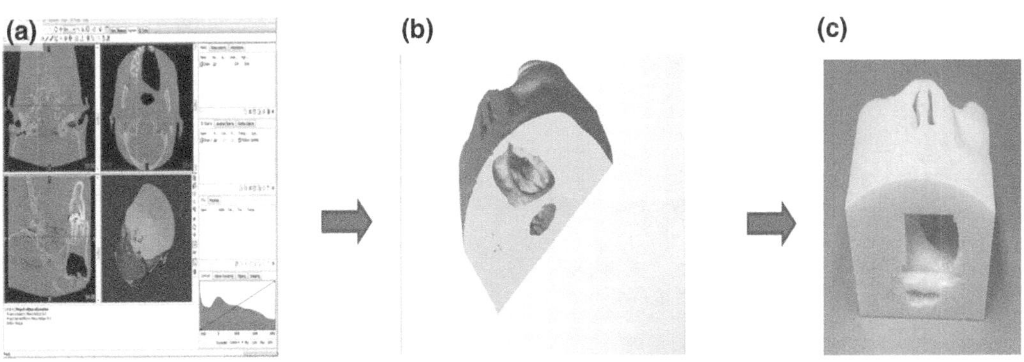

Figure 3.15 (a) Patient imaging from CT showing relevant anatomy. (b) Segmented digital model. (c) 3D-printed model used to create prosthesis.

both sides of the fracture. The result of this process is a bone segment that is fused together and maintains its orientation. Typically, the fixation plate is chosen and bent during the procedure. Depending upon the complexity of the case and the experience of the practitioner, the surgical time taken to bend these plates accurately can be in the tens of minutes. While this doesn't seem like much, every minute of anesthesia time for the patient, operating room time, and surgical team time represents both a clinical and financial cost for the patient and hospital. An existing CT scan of the patient was used to segment and reproduce the broken mandible and it was then digitally manipulated to bring the broken segments back into their intended locations/alignment. A 3D-printed model (Figure 3.16) was made that allowed the clinicians to pre-bend the fixation plate, taking as much time as they needed without the pressure of having to do this with an anesthetized patient. The fixation plate was accurately positioned, bent, and sterilized for clinical use on the patient. This is an excellent use of 3D printing technology to reduce operating room time and potentially improve outcomes. Furthermore, these applications can use the most cost-effective printers to accomplish this as there are no requirements for advanced material properties.

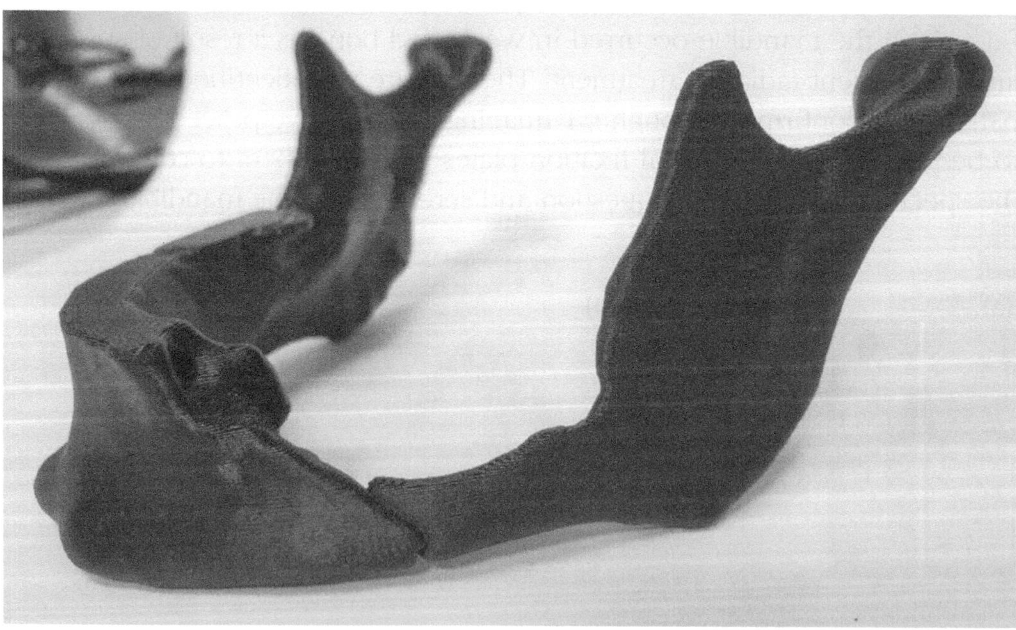

Figure 3.16 3D-printed mandibular model for pre-bending of fixation plates. The mandible is repositioned, but the fracture is still visible.

Research and Development

This category describes the multitude of applications which fall under a hospital's R&D or training activities. These can be devices made for training, testing, experimentation, prototyping, and so on. For research or teaching hospitals, 3D printing is proving to be a critical technology which has become an integral part of the development process. The range of applications in this space is quite broad, from simple mounts/adapters/jigs in experimental setups to full anatomical models used to train nurses, residents, and fellows.

An example of an experimental setup print is shown in Figure 3.17. Here, a test was needed to quantify the effect of a source light emitter's angular position on the signal intensity seen by the imaging system. A custom rotating mount was made which holds the light source in place and can be rotated to specific angular positions for testing [16]. Multi-material PolyJet technology was used here to allow integral markings in the print.

As a cancer research and treatment center, our institution has a large focus on development work for radiation therapy and testing. We create many custom radiation shields (patient- or application-specific) which can be 3D-printed hollow and then filled with appropriate shielding material. Shielding material can be, for example, small metal spheres such as stainless steel, copper, or tungsten, or alternatively filled with Cerrobend (a low-melt temperature lead alloy). These can be, for example, a small patient-specific eye shield or a relatively large enclosure for sensitive electronics. These prints allow incorporation of shielding along with other features needed in a custom testing setup.

Another application which is advancing quickly within the Radiation Oncology and Medical Physics space is the 3D printing of imaging phantoms. Phantoms are used regularly in radiation oncology for testing, QA, and plan checking. There are materials which can be 3D-printed that have variable properties when imaged by CT and/or MRI, and these materials can be used to create custom-configured CT or MR phantoms. Commercially available phantoms are often very expensive and are available in a limited set of configurations. 3D printing allows custom phantoms to be made to suit exactly what is needed for a given test.

For teaching hospitals, one of the leading applications for 3D printing is in creating training models for medical staff. Even with existing technologies, models can be used to simulate various procedures. Rapid advancements in material properties are allowing increasingly realistic models which can

Figure 3.17 Rotating mount for testing the angular dependence of the light emitter.

now supplant the use of cadavers, animal models, or practicing directly on patients. For example, 3D-printed models can be used to simulate a complex airway (e.g., specific tumor physiology) and allow prior practice of intubation or bronchoscopy. Realistic tissue materials provide tactile feedback and allow development of approach techniques to handle these complex scenarios. As of early 2025, nearly every field of surgical practice is currently exploring applications for these kinds of models.

Future Use Cases

3D printing technology is a rapidly evolving field with significant investment into both refining existing processes and innovating new technologies. Areas like bioprinting, electronics printing, MEMS (micro electromechanical systems) sensors and actuators, life-like materials, and entire printed systems that combine some or all of these technologies are being investigated and are at varying stages of maturity. In the materials space, new materials with focused properties that take advantage of existing printing processes are introduced into the market on a regular basis.

In the 3D-printed electronics space, research is moving from simply printing conductive traces for connectivity between different components to printing complete embedded systems and components down to the transistor level. Printable displays and sensors are some of the targeted areas for development that will take advantage of the unique abilities that 3D printing affords. It is allowing designers to think outside of their traditional processes and limitations and potentially allow for more robust designs and applications.

On the medical side of this evolution, we are seeing new materials introduced that mimic the look, feel, and response of actual biologic materials. This is enabling the industry to develop a new range of anatomically accurate models to be used for education, simulation, and more effective training. One of the ultimate goals in the field is to print living biological tissue which can repair or even replace human anatomy; this technology may eventually be able to provide complex systems, including functioning organs.

Conclusion

3D printing technology has rapidly gained adoption in the hospital environment due to its ability to create patient-specific medical devices, custom surgical tools, anatomical models, implants, research tools, and other applications. The flexibility of 3D printing provides a wide range of options for clinical applications, enabling engineers, radiologists, and other physicians to collaborate and create bespoke, on-demand solutions. With the increasing adoption of this technology, a complete workflow for the delivery of printed products to the point of care has been established, creating new clinical opportunities and allowing for quick responses to unforeseen needs, as demonstrated during the COVID-19 pandemic.

Implementing a clinical 3D printing service in a hospital requires establishing proper facilities, legal and medical approvals, proper handling of materials, documentation, QA, and funding/billing. Centralizing the 3D printing operations can offer advantages in terms of cost-effectiveness, higher utilization rates, and more efficient use of staffing and technical resources. The specific mix of printer technologies and their associated capabilities should be chosen to provide a combination of targeted capabilities for clearly identified clinical use models as well as flexibility to allow for the broadest possible use cases.

Overall, the adoption of 3D printing technology in hospitals is becoming an increasingly necessary infrastructure to provide advanced healthcare and add resiliency to clinical operations. As the technology continues to evolve, we can expect to see even more innovative and practical clinical applications emerging in the years to come.

References

1. S. Hellman, P. Frisch and P. Booth, "Medical devices: Responding to a crisis with 3D printing, design, engineering, and production," in *Leveraging Technology as a Response to the COVID Pandemic*, Taylor & Francis, 2022.
2. S. Bastawrous, L. Wu, P. Liacouras, D. Levin, M. Ahmed, B. Strzelecki, F. Amendola, L. Coburn and B. Ripley, "Establishing 3D printing at the point of care: Basic principles and tools for success," *Radiographics*, vol. 42, no. 2, pp. 451–468, 2022.
3. K. Shine, L. Schlegel, M. Ho, K. Boyd and R. Pugiliese, "From the ground up: understanding the developing infrastructure and resources of 3D printing facilities in hospital-based settings," *3D Printing in Medicine*, vol. 8, no. 21, 2022.
4. J. Agueda, Q. Chen, R. Maalihan, J. Ren, I. da Silva, N. Dugos, E. Caldona and R. Advincula, "3D printing of biomedically relevant polymer materials and biocompatibility," *MRS Communications*, vol. 11, pp. 197–212, 2021.
5. Food and Drug Administration, *Discussion paper: 3D printing medical devices at the point of care*, 2022.
6. Z. Rahman, S. Barakh Ali, T. Ozkan, N. Charoo, I. Reddy and M. Khan, "Additive manufacturing with 3D printing: Progress from bench to bedside," *AAPS Journal*, vol. 20, no. 6, 2018.
7. V. Paramasivam, V. Sindhu, G. Singh and S. Santhanakrishnan, "3D printing of human anatomical models for preoperative surgical planning," *Procedia Manufacturing*, vol. 48, pp. 684–690, 2020.
8. N. Wake, J. Wysock, M. Bjurlin, H. Chandarana and W. Huang, "'Pin the tumor on the kidney:' An evaluation of how surgeons translate CT and MRI data to 3D models," *Urology*, vol. 131, pp. 255–261, 2019.
9. P. Ravi, M. Burch, S. Farahani, L. Chepelev, D. Yang, A. Ali, J. Joyce, N. Lawera, J. Stringer, J. Morris, D. Ballard, K. Wang, M. Mahoney, S. Kondor and F. Rybicki, "Utility and costs during the initial year of 3-D printing in an academic hospital," *Journal of the American College of Radiology*, 2022 [In Press].
10. L. Diment, M. Thompson and J. Bergmann, "Clinical efficacy and effectiveness of 3D printing: a systematic review," *BMJ Open*, vol. 7, no. 12, 2017.
11. D. Ballard, P. Mills, R. Duszak, J. Weisman, F. Rybicki and P. Woodard, "Medical 3D printing cost-savings in orthopedic and maxillofacial surgery: Cost analysis of operating room time saved with 3D printed anatomic models and surgical guides," *Academic Radiology*, vol. 27, no. 8, pp. 1103–1113, 2020.

12. N. Wake, T. Rude, S. Kang, M. Sifleman, J. Borin, D. Sodickson, W. Huang and H. Chadarana, "3D printed renal cancer models derived from MRI data: Application in pre-surgical planning," *Abdominal Radiology*, vol. 42, no. 5, pp. 1501–1509, 2017.
13. J. Meyer-Szary, M. Luis, S. Mikulski, A. Patel, F. Schulz, D. Tretiakow, J. Fercho, K. Jaguszewska, M. Frankiewicz, E. Powlowska, R. Targonski, L. Szarpak, K. Dadela, R. Sabiniewicz and J. Kwiatkowska, "The role of 3D printing in planning complex medical procedures and training of medical professionals—Cross-sectional multispecialty review," *International Journal of Environmental Research and Public Health*, vol. 19, no. 6, 2022.
14. R. Moore, W. Wallen, K. Riggs and D. Morales, "Three-dimensional printing in surgical planning: A case of aortopulmonary window with interrupted aortic arch," *Annals of Pediatric Cardiology*, vol. 11, no. 2, pp. 201–203, 2018.
15. A. Inserra, L. Borro, M. Spada, S. Frediani and A. Secinaro, "Advanced 3D 'modeling' and 'printing' for the surgical planning of a successful case of thoraco-omphalopagus conjoined twins separation," *Frontiers in Physiology*, vol. 11, 2020.
16. E. Pratt, M. Skubal, B. McLarney, P. Causa-Andrieu, S. Das, P. Sawan, A. Araji, C. Riedl, K. Vyas, D. Tuch and J. Grimm, "Prospective testing of clinical Cerenkov luminescence imaging against standard-of-care nuclear imaging for tumour location," *Nature Biomedical Engineering*, vol. 6, no. 5, pp. 559–568, 2022.

Section 4

Technology Management within the Intelligent Hospital

Introduction

Rise of the Intelligent Hospital has evolved through a continuous development and implementation of advanced technologies including medical devices, clinical applications, network infrastructure, and wireless technologies supporting large-scale integrations between these devices and clinical applications and exploiting a variety of enabling and diverse IoT technologies. These solutions enable and focus on the creation of a seamless patient care data environment which enables the rapid exchange of patient specific actionable information throughout the continuum of care. This dynamic environment provides the ability to develop and deploy state-of-the-art complex systems which directly and positively improve patient outcomes, enhance patient safety, and improve overall satisfaction. Dynamic institutions maintain their edge by being at the forefront of research, innovation, and technological advances, rapidly integrating and transitioning these diverse areas into diagnostics, treatment, and the overall delivery of patient care. These enhanced solutions provide high-quality validated actionable data across all aspects of the patient care environment, which is key to the development of the Intelligent Hospital.

As discussed throughout, the fundamental objectives of the Intelligent Hospital focus on establishing a seamless and standardized patient care environment across the entire continuum of care including the hospital,

 DOI: 10.4324/9781003414568-4

ambulatory care, regional network, and all supporting locations enabling and providing a broad scope of functionality as outlined:

- Increased Data Accuracy & Availability
  - Electronic Medical Records
  - Shared Medical Device Interconnectivity and Integration
  - Global Visualization of the Patient/Overall Care Unit Information
- Optimize Rapid Information Dissemination and Response
  - Physiological Data
  - Alarm Event Management
  - Advanced Communication Platforms (PDA, iPhone, etc.)
  - Location Content
- Maximize Enhanced Error Checking and Validation
  - Identification and Association
  - Location Content
  - Correct Information
- Drive the Optimization of Workflow and Resource Management

These objectives require a continued focus on integration, design, development, deployment, and optimization, all of which are supported through an ongoing review process and advanced management strategy ensuring these diverse set of technologies, advanced medical devices, clinical applications, and various IoT devices all integrate to seamlessly share and deliver information across the continuum of care. These integrated technologies provide the tools and support the mechanisms necessary to establish and validate connectivity, provide unique identification, and enable critical communications, association, and location tracking defining and driving the delivery of accurate information to enhance the patient care process and optimize the effectiveness and efficiency across the entire patient care environment.

To achieve the desired planned objectives and advanced operability, new management and support strategies need to be developed and implemented, including strategic planning, technology management, for validation of operations, and reengineering solutions to ensure operational integrity, emergency management, restoration, and recovery. The specific issues as outlined in Table 4.1 can vary as a function of the institution and the overall design strategy potentially impacting both the types of issues the magnitude.

Table 4.1 Management of Complex Systems and Operations

- Strategic Planning
- Technology Complexity
  - Big Data
  - Patient Care Environment
  - Infrastructure
  - Complex Medical Systems
- Reengineering Technology Support Models
  - Biomedical Engineering Organizational Changes
    - Synergies with Other Technology Groups
- New Operations and Tools
  - Dynamic Simulation Laboratory
    - Upgrades and Testing
    - New Devices and Technologies
    - Software Upgrades
- Cybersecurity Issues
  - Cybersecurity Assessment
  - Real-Time Vulnerability and Threat Monitoring
  - Response and Recovery Processes and Procedures
- Developing Enhanced and Optimized Workflows
- Staff Education and Training
- Emergency Management
  - Case Study: Emergency Management during COVID-19

Strategic Planning

The design, development, implementation, and continuous technology upgrade supporting the Intelligent Hospital require a clear vision and well-defined strategic plan. This plan clearly defines the overall objectives, which may vary and need adjustment over time, and specifically identifies the required technologies, integrations, deliverables, and data flows needed to maintain and manage the seamless data flow within the environment of care. Organizations which are fortunate enough to design and build new healthcare facilities from the ground up can design and implement broad based institutional solutions, which capitalize on standardization of medical devices, clinical applications, and enabling technologies which then drive the design of robust networks and wireless infrastructure to support the massive data exchange, flow, and archival. This standardization and planning process enables support, maintenance, and technology management solutions to be defined and established as part of the overall planning, design, development, and budgeting phases.

However, most organizations do not frequently have the opportunity to design the Intelligent Hospital from scratch and require that the design, development, and implementation phases of the Intelligent Hospital be created piecemeal based on a comprehensive and dynamic transition plan. The transition strategy of how and when to incorporate new, evolving, and enabling technologies is driven by many factors including the budgetary processes and developing comprehensive plans of how to integrate existing legacy systems supporting the inclusion of existing workflows into the evolving Intelligent Hospital. To ensure a successful evolution, there must be an ongoing focus on how to support, maintain, and manage this highly integrated and dynamic hospital environment.

These challenges require new and enhanced management and support synergies as well as new tools to ensure optimum utilization and operational integrity of these complex systems. These synergies must transcend the traditional department boundaries to define and address the varying skill sets necessary to support this dynamic environment. The responsibilities of Biomedical Engineering, Information Technologies, Facilities Management, as well as Clinical Departments all play a role in ensuring operational integrity and patient safety. How various hospitals implement these synergies directly impact the overall level of success.

Complexity of the Patient Care Environment

Review and analysis of the patient care areas, as illustrated in Figure 4.1, indicates a variety of configurations which enhance both medical systems and clinical applications user interfaces as a function of the care environment. Figure 4.2 illustrates a simulation training ICU critical care area representing an extremely complex patient care environment that includes medical devices with multiple interfaces with a variety of clinical software applications, which also include visualization via dashboards and hand held devices. However, each of the areas consists of a combination of devices, systems, and technologies, some standardized or common throughout the various hospital care environments and others which include devices and systems unique to the operational functionality of the care environment. All the environments are supported through a robust networked environment, both wired and wireless along with a variety of dedicated frequency antenna systems to support RFID and IoT devices used to gather and deliver clinical and demographic data, imaging, and alerts and alarms. This environment enables rapid data distribution, assessment and decision support, staff to staff communications, and unique visualizations of patient information shared between various clinical groups and physicians in the Intelligent Hospital. This care environment enables and positively impacts clinical response, treatment, patient outcome, and patient safety.

Figure 4.1 Patient care environments.

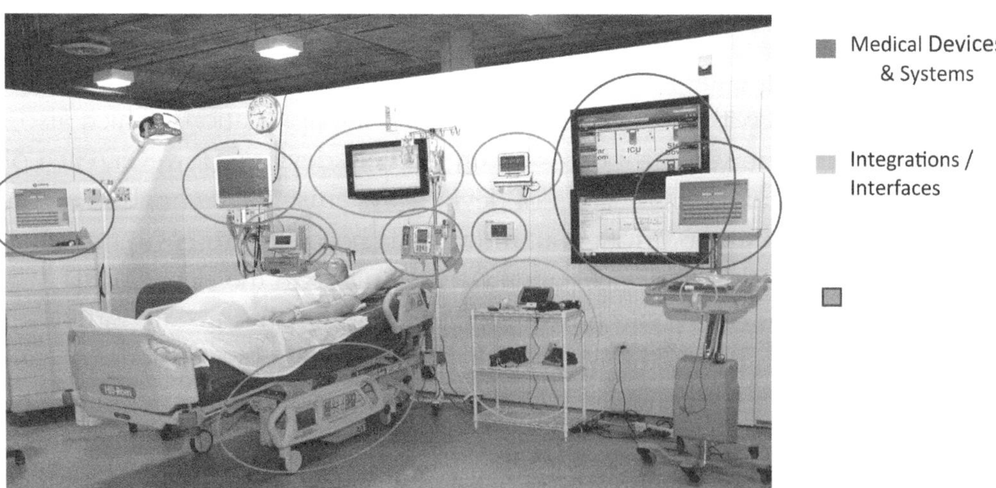

Figure 4.2 Simulation/education ICU laboratory.

The specific types of medical devices/systems, clinical applications, and enabling technologies, as well as their specific configurations, vary as a function of room type and care modality. The Intelligent Hospital care environment can be broadly defined into several distinct areas, specialty or treatment areas, in-patient care, ambulatory care, supporting clinical and management areas, and telemedicine or home care.

Specialty/Treatment Areas

Specialized treatment or care areas typically include highly specialized medical devices as well as a standardized set of medical devices and clinical applications, IoT devices, locating technologies (RTLS, etc., environmental monitoring systems used throughout the Intelligent Hospital. In many cases, the specialty and treatment areas outlined are frequently supported through service support agreements or contracts, most commonly by OEM (original equipment manufacturer) directly. However, within the Intelligent Hospital where devices and systems integrate other systems, applications, and enabling technologies, these are jointly managed providing a shared risk approach directly within internal staff that is highly trained on the operation, management, and evaluation of the operational integrity of these specialized devices. These commonly include:

- Surgical Suites
  - Operating Room Platforms
    - Endoscopic Procedures
    - Surgical Laser
    - Robotic Surgical Systems
  - Labor and Delivery
- Radiation Therapy
  - Linear Accelerators
  - Proton Systems
  - Brachytherapy Systems
- Imaging/Radiology
  - Hybrid Systems/CT/MR/PET
  - Interventional Radiology
- Clinical Laboratories

In-Patient and Ambulatory Care

Within the in-patient and ambulatory care patient care and management areas, medical devices, clinical applications, and supporting technologies are increasingly standardized to provide broad-based institutional systems and solutions which include but are not limited to the following:

- **Medical Devices**
  - Instrumented Patient Bed
  - Nurse Call
  - Connectivity Interface
  - Infusion System (Pump) (1–3 pumps)
  - Physiological Monitoring
    - Telemetry
  - Vital Sign Monitor/Physiological Monitoring
    - Blood Pressure
    - SpO_2
  - Ventilator
  - Defibrillator
  - Transport (Wheelchairs and Stretchers)
- **Clinical Applications**
  - Clinical Workstation and Dashboards (Wall Mounted or Mobile Unit)
  - Electronic Medical Record
  - PACS Image Archival System
  - Physician Order System
  - Alarm Management
  - Scheduling Systems
  - Admit and Discharge
- **Support System Applications**
  - Sample Collection
  - Medication Administration
    - Medication Carts, Clinical Laboratory
  - Supply Management
    - RFID Cabinets
  - Inventory Management
  - Admit and Discharge
- **IoT Devices**
  - Mobile Devices (Laptop, Tablet, iPad, iPhone)
  - Bar Code/QR Readers/Scanners
  - Location and Tracking Solutions (RTLS/RFID, BLE)

- **Visualization Dashboards**
  - Physiological Data (Wall of Knowledge)
  - Demographics
  - Location and Tracking Solutions
  - Alarm Notification
  - Cybersecurity Alerts
- **Environmental Monitoring**
  - Hand Hygiene
  - Light and Sound Monitor
  - Particulate
  - Temperature
  - Humidity

Managing, maintaining, and supporting the operational integrity within these patient environments require continuous education to maintain a high level of proficiency to operate the devices and provide troubleshooting and maintenance to ensure functionality and regulatory compliance. These complex systems will also require new types of evaluation tools and a test environment (Simulation Laboratory) to enable the operational validation, maintenance, upgrades, and expansions to be tested and evaluated. This will be further discussed and detailed later in the section.

Regulatory Compliance

The complexity of the evolving Intelligent Hospital infrastructure, medical systems, and enabling technologies within the patient care environment will establish new guidelines for compliance, operational verification, and validation as part of the regulatory process, including but not limited to the following issues.

Management of Big Data

Fundamental to the design of the Intelligent Hospital is the creation and establishment of a robust and seamless shared data environment extracting data from a large quantity of complex networked medical systems, where each system consists of integrated medical devices, clinical applications, and supporting IoT devices. Each component of this complex medical system can generate large volumes of data. This advanced patient care environment requires a robust methodology to manage these large volumes of data consisting of different types of systems, coming from multiple sources with potentially different formats. This data needs to be uniquely identified, associated, time synchronized, sorted, aggregated, collated, filtered, and processed based on a well-defined dynamic relationship to support the clinical data flow and workflow.

The data management, flow, and delivery of this data requires a consistent and reliable patient centric association process [1], associating the patient with the care providers, with the medical devices, pharmaceuticals, supplies, etc., as illustrated in Figure 4.3.

The data itself covers a broad spectrum of data types, including clinical patient data, physiological parameters, alarms/alerts, patient demographics, and analytics, potentially coupled with other operational data as summarized below.

- Clinical Data
  - Physiological Parameters (ECG, SpO_2, Temperature, etc.)
  - Physician Orders
  - Laboratory Information
  - Imaging

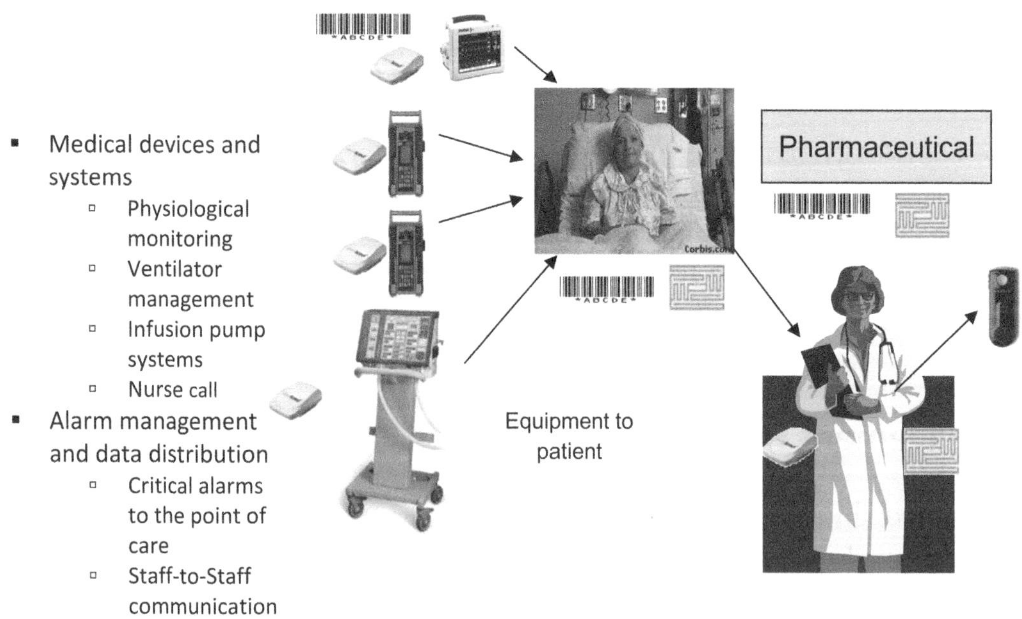

Figure 4.3 Patient-centric identification and association.

- Alarm Notification and Management
  - Physiological Alarms
    - Arrhythmia
    - Rate, Range
  - Device Operations and Cybersecurity
- Real-Time Location Content
  - Workflow Metrics
- Environmental Data
  - Temperature and Humidity
  - Light and Sound
  - Fire Protection
  - Radiation
- Resource Availability
  - Staff
  - Supply Chain
    - Equipment and Supplies
- Administrative/Management Information
  - Business Intelligence

Example: Case Study: Magnitude Physiological Monitor Alarms

The following example highlights the alarm volumes associated with physiological monitoring. The Intelligent Hospital enables and provides the capability to process aggregated groups of data and associated information in many ways and deliver this information to a broad base of devices (handheld devices, dashboards, etc.) and users, including clinical staff, physicians, nursing, administration, Biomedical Engineering, supply managers, etc. As an example, Figure 4.4 illustrates the alarm output generated from a telemetry-based physiological monitoring system in a 42-bed cardiac telemetry unit. In order to manage and provide the optimized response to this extensive alarm volume, the Intelligent Hospital utilizes advanced analytics providing a variety of processing algorithms, including alarm prioritization, filtering, and thresholding methodologies to reduce, prioritize, and manage the alarm volumes. Specific alarm algorithms can exploit patterns and identification techniques to formulate and identify trends which can be translated and correlated to specific physiological events indicating changes to a patient's risk and need for attention. These trends and patterns enable prioritized rapid response which can directly alter and effectively impact patient outcomes. By comparison, the typical

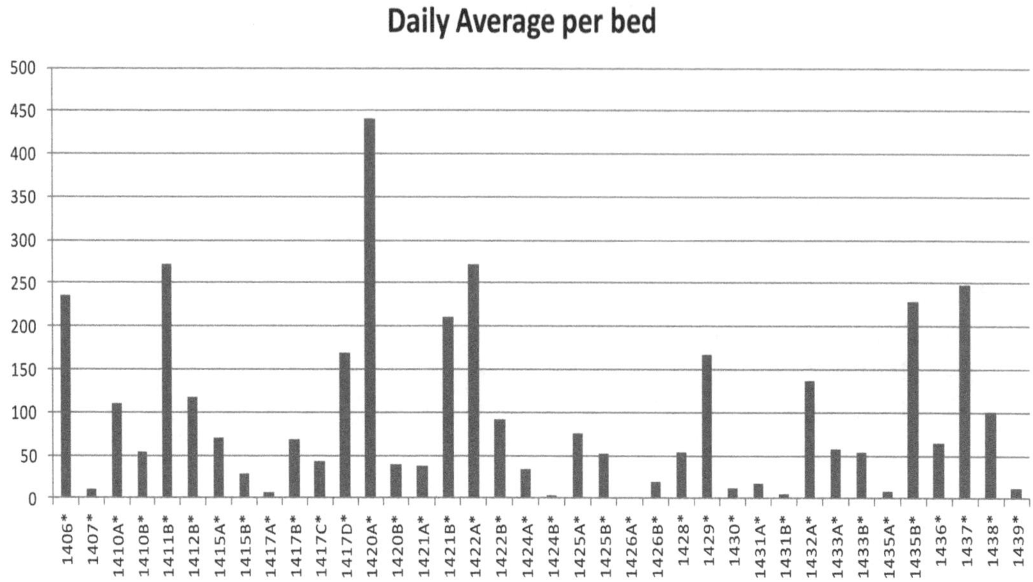

Figure 4.4 Magnitude (volume) physiological monitoring alarms.

quantity of alarms from a critical care area, such as an ICU or cardiac care unit (CCU), can be ten times greater than the outlined example. Specific alarms such as large numbers of PVCs may have minimal impact on the patient, however, can drastically increase alarm volumes. A challenge to the clinical staff is to prioritize and define the workflows necessary to optimally address or manage these large volumes of information and determine and provide the appropriate clinical responses.

Evolving Advanced Biomedical Engineering Operation Requirements

As we outline and discuss the extended technology management roles of Biomedical Engineering in the Intelligent Hospital, several institutional requirements are clearly defined. Biomedical Engineering must support capabilities to provide highly robust and interactive capabilities to support the following institutional operations:

- Emergency Management
- Medical Device Cybersecurity
- Clinical Simulation and Advanced Technology Evaluation

These functions have become very interleaved and represent institution-wide issues of concern. Biomedical Engineering must develop the synergies and the processes to evaluate, test, and drive processes and procedures to minimize the impact of new and evolving threats.

In 2019, as the Coronavirus pandemic hit the world, hospitals were forced to quickly develop and manage operations and technologies to address significant patient loads with unknown treatments and implementation of processes to best manage and support this unexpected and unpredictable virus. What we learned during this period provided the baseline of next-generation emergency management protocols, processes, and preparedness.

Emergency Management Preparedness

Emergency management or disaster management is a critical function within a hospital which creates the framework and defines processes to reduce vulnerability to hazards and respond and deal with disasters directly impacting hospital operations. These include natural disasters, such as earthquakes, floods, fires, pandemics, as well as disasters resulting from terrorist attacks and even cybersecurity. Any event that has the potential of disrupting hospital operations and limiting the ability of the healthcare facility to deliver patient care needs to be considered and addressed by the hospital emergency plan which focuses on the response to restoring operations. With hospitals' increased dependence on technology, power, interconnectivity, network operations, wireless technology, etc., natural and emergency events can quickly and directly impact, disrupt, and negatively impact the patient safety and care process and environment. The design of the Intelligent Hospital must focus and establishing backup and recovery processes to enable rapid recovery minimizing the impacts of these events.

The care process can be impacted through many scenarios, which are discussed and covered in many books and throughout the literature, including fire, loss of power, loss of connectivity, networking, wireless, etc. Any fully functional operation can be directly impacted by the presence of unexpected large volumes of patients producing driving scenarios which exceed an institution's ability to provide care to these patients. The scenario results in bed and medical device shortages as well as issues with supplies, medications, and even food. The following section discusses the specific issues associated with the unexpected COVID-19 pandemic event.

Case Study: Impact of Corona Virus (COVID-19) and Response

An unexpected emergency event impacted all medical facilities throughout the world starting in December 2019. The novel coronavirus, designated SARS-CoV-2, caused an international outbreak of respiratory disease identified as the novel **CO**rona **VI**rus **D**isease 20**19** or COVID-19 [2]. By February 11, 2020, the World Health Organization (WHO) declared the disease a pandemic [3]. By April 2020, more than 1.2 million cases were reported worldwide, resulting in more than 72,000 deaths [4]. As a result, in early 2020 the world healthcare systems were experiencing a global pandemic event, which over the course of 2020 and continuing through 2021 spread across the globe, quickly impacting all healthcare systems which were unprepared to deal with this type of large-scale event. Over the course of the pandemic, the world has experienced multiple peaks in the number of infections or transmission cases of the disease, resulting in 276 million cases (Figure 4.5) and 5.34 million deaths (Figure 4.6) worldwide by March 22, 2022 [5]. The resulting influx of patients presenting at hospitals coupled with the large volume of admissions and subsequent patient deaths rapidly exceeded the capacity of Hospital's and the overall healthcare systems capacity in terms of available beds, medical equipment, protective devices, supplies, overall resources, and staffing.

Coronavirus (COVID-19)

Coronavirus (COVID-19) is a novel virus, which means that humans do not have a natural immunity or ability to fight it off with their immune systems. The first known case was identified in Wuhan, China, in December 2019 [5]. COVID-19 is a highly contagious airborne virus that primarily enters the body through the eyes, nose, or mouth and progresses into the lungs, causing severe acute respiratory syndrome. Transmission occurs when people are exposed to and breathe air contaminated by droplets and small airborne particles. The risk of breathing in these particles is highest when people are in proximity, but they can also be inhaled over longer distances, especially indoors. People can remain contagious for up to 20 days; however, experts believe that the time from exposure to symptom onset, also known as the incubation period, is 2–14 days. However, symptoms typically appear within 4–5 days of exposure [6]. It is important to note that not all cases exhibit symptoms and that the virus can be spread even if a person does not develop any symptoms [7].

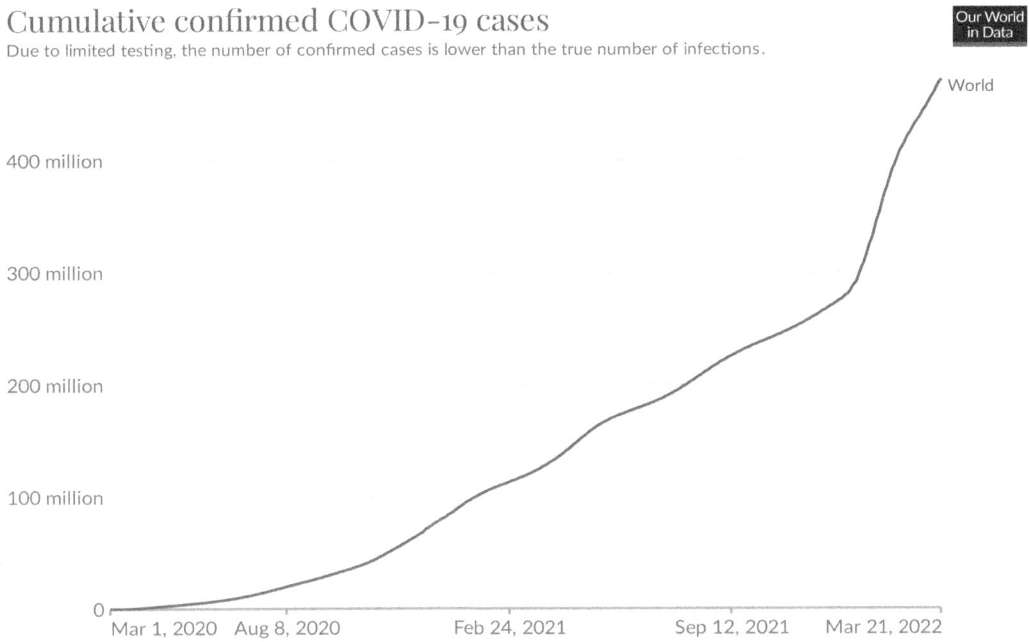

Figure 4.5 Cumulative confirmed COVID-19 cases. (John Hopkins University CSSE COVID-19 data, March 22, 2022).

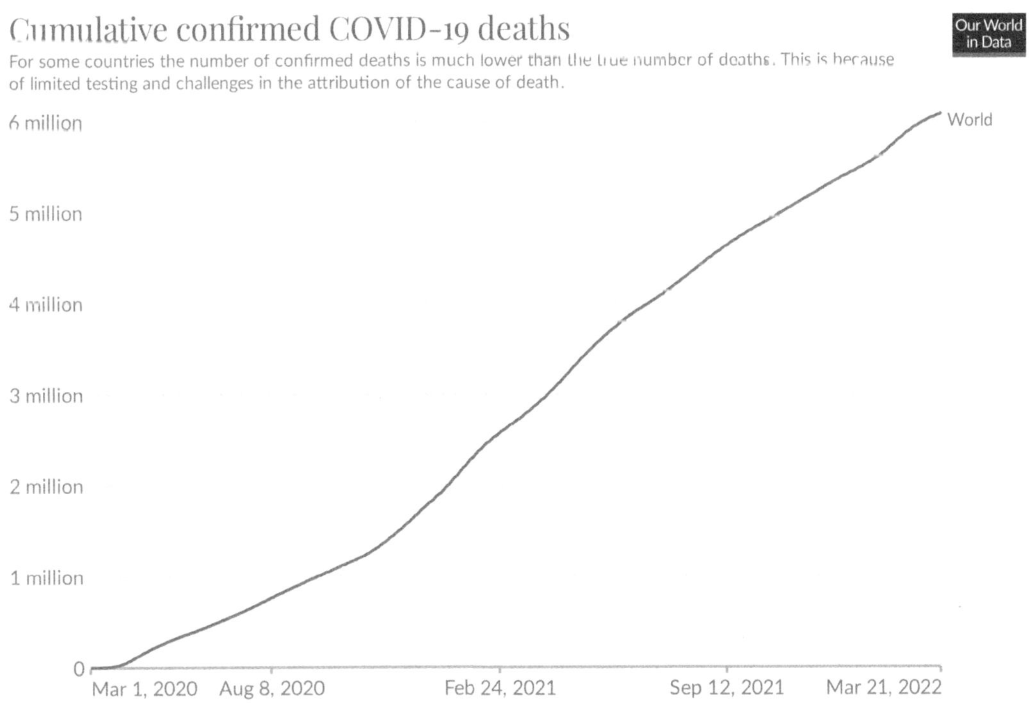

Figure 4.6 Cumulative deaths COVID-19 worldwide. (John Hopkins University CSSE COVID-19 data, March 22, 2022).

COVID-19 Surges

To understand the impact on healthcare systems, it is important to establish metrics on the rapid onset and spread of the virus. Review of the statistics and data associated with surges provides a clear visualization of the stress on hospitals and healthcare organizations. The following section summarizes and highlights the data in terms of the world, the United States, New York State, and New York City. There are many sources of statistical information which are available from a variety of government agencies, universities, the CDC, WHO, etc. Many of these sources provide data download capability enabling the COVID data to be analyzed from many perspectives [4, 8–10]. The impact of the virus to date (as of March 22, 2022) can be summarized as follows:

Country (Locations)	*Total Number of Cases*	*Deaths*
World	470,737,068	6,077,582
United States	79,734,788	972,361
New York State	4,948,160	67,203
New York City	2,290,000	40,020

In addition, the data used to identify the magnitude of the infections and consequent deaths needs to be analyzed in terms of surge numbers and hospital impact. For this review and analysis, let's examine the impact of the initial patient spike or surge which resulted in an unprecedentedly large number of patients presenting at hospitals for testing, diagnostics, treatment, and admission. Figure 4.7 shows an overview of the number of daily admissions or hospitalizations in New York City over the course of the pandemic (Feb 2020–June 2021).

Similarly, Figure 4.8 details the resulting number of daily deaths due to COVID-19 over the same time period. As shown, the initial spike occurred in the end of February through the May 2020 time frame followed by a secondary lower level, but longer duration peak started in the fall of 2020 and continued through the spring of 2021. In the first quarter of 2020, New York City became the epicenter of the coronavirus pandemic in the United States, putting historic pressure on the City's world-renowned healthcare institutions as the number of confirmed cases in the area grew exponentially. New York became the model of what other cities and areas of the United States would experience as the virus spread across the country.

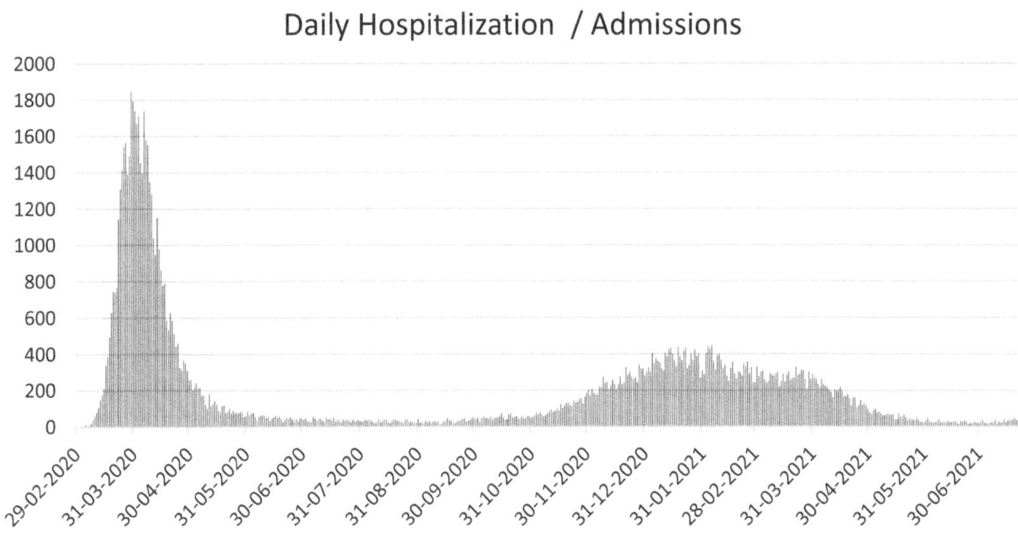

Figure 4.7 Daily hospitalizations/admissions between February 2020 and June 2021.

As we examine the impact of the hospitalizations based on the initial surge data, we can review the statistics and extrapolate key trends creating the overload conditions within the health systems. COVID-19 patients exhibited lengthy hospital stays with a median length of stay (LOS) ranging from 4 to 21 days, with 4–19 days typical for ICU patients [11].

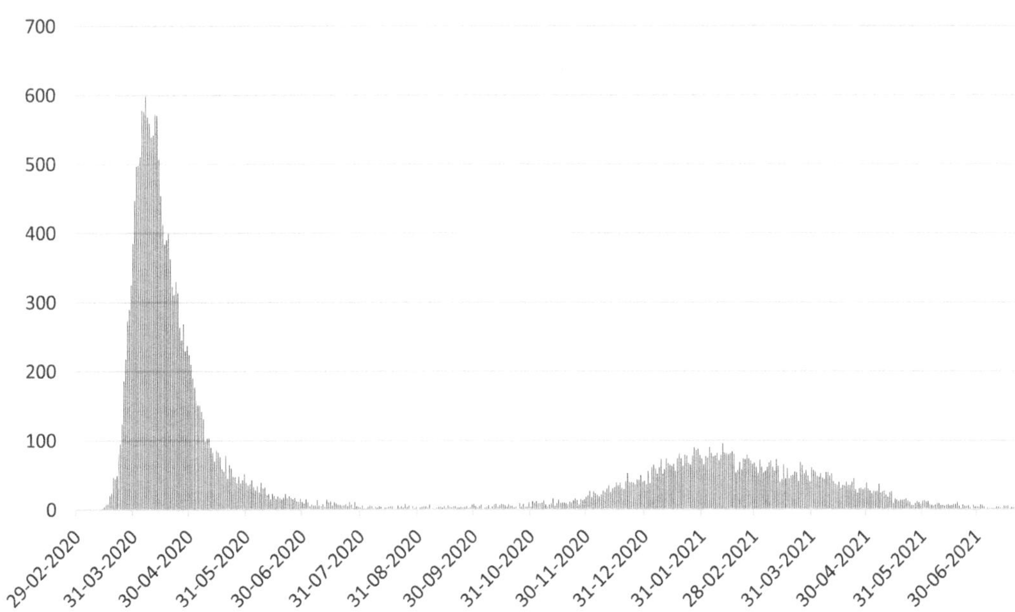

Figure 4.8 Daily COVID deaths (February 2020–June 2021).

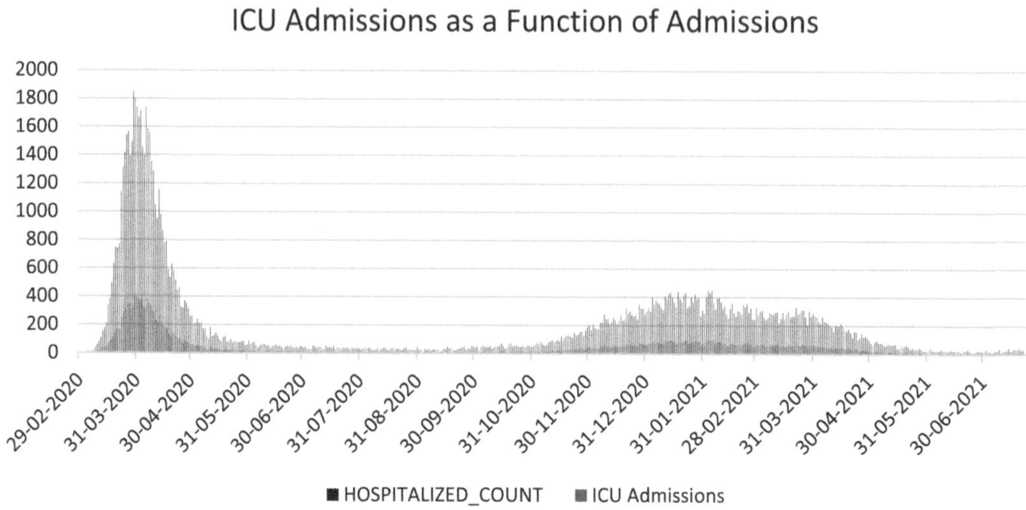

Figure 4.9 ICU admissions (bed requirements) as a function of admissions.

As reported by the US News based on a dataset published in the Lancet Journal on May 19, 2020, NYC patient hospitalizations resulted in 22% of all admissions requiring critical care within an ICU setting, of which 79% of these ICU admissions required ventilation. These factors directly contributed to the increasing need and urgency for ICU or critical care level facilities. These facilities include critical care beds, isolation areas, and advanced medical devices and systems, including physiological monitoring, ventilation, and infusion systems. Based on the statistic that 22% of all the admissions required critical care, we can then project the daily demand for ICU beds as represented in Figure 4.9, indicating that at the peak approximately 400 additional ICU beds are required daily to support the admission.

Impacts on Bed and ICU Capacity

According to the American Hospital Association (AHA), there are a total of 6,146 hospital beds housed in 62 hospitals located throughout the New York City Metro area with a combined capacity of 15,659 beds which includes 2,111 ICU level beds [12]. To fully understand the impact of the COVID patient demand for beds, especially ICU level beds, one needs to examine the typical ICU bed availability and the LOS or utilization of the ICU beds.

For most level 1 trauma centers and tertiary care facilities, intensive care units typically operated at 80–90% capacity even before the COVID-19 pandemic hit [13]. As reported by FOX News on Dec 31, 2020, summarizing a New York Times report, highlighted a study published by the Society of Critical Care Medicine which supported the statistics of typical ICU occupancy which ranged from 57.4% to 82.1% [14, 15].

If we examine bed utilization as a function of patient LOS to determine bed availability, we can develop a broader view of how resources quickly diminish as a function of the increased LOS associated with COVID patients. Patient LOS varies as a function of the type or acuity of the illness and treatment protocols and procedures. The average LOS in a hospital is commonly used as a measure or gauge of the efficiency of a healthcare facility. The national average for a hospital stay is 4.5 days, according to the Agency for Healthcare Research and Quality [16]. The average ICU LOS is 3.3 days, and for every day spent in an ICU bed, the average patient spends an additional 1.5 days in a non-ICU bed [17]. In the case of COVID-19 patients, hospitalization means the LOS can be estimated to be between 15.8 and 20.1 days [18]. The impact of the increased LOS limits a hospital's bed turnaround rate rapidly reducing bed availability.

The combined impact of increased number of patients requiring ICU levels bed coupled with the increase length of stay quickly impacts and exceeds the capabilities of most hospitals (Figure 4.10).

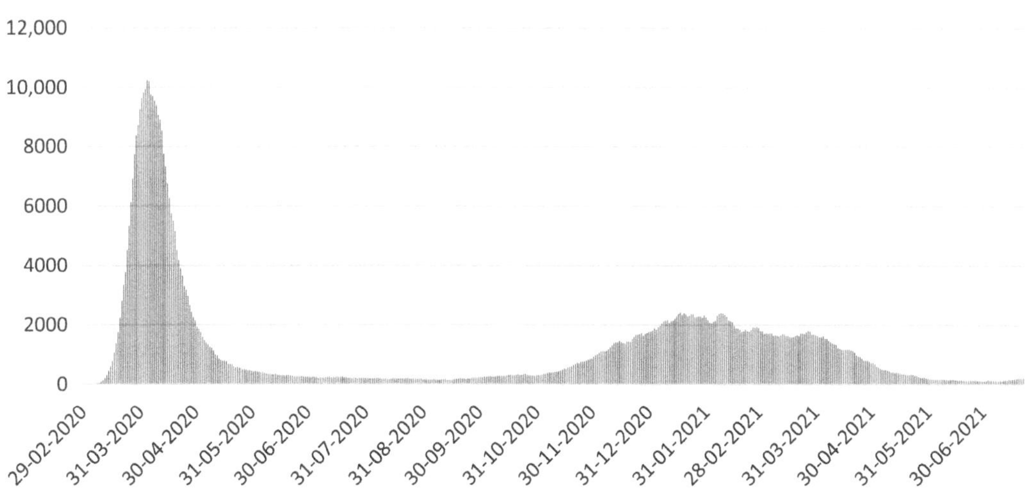

Figure 4.10 Required bed capacity.

Hospital Impact and Expansions Efforts

In response to this magnitude of patient admissions, Hospitals across New York started to erect temporary facilities, tents, reworked space and doubling up rooms where possible, in an effort to add beds and increase bed availability to support the increasing number of confirmed coronavirus cases. In March of 2020, Governor Andrew Cuomo issued Executive Order No. 202.5, which allows the state to rapidly increase hospital capacity to prepare the NYS healthcare system to handle the potential influx of patients suffering from COVID-19. To accommodate the anticipated increase in admissions, Governor Cuomo mandated the NYS Department of Health to suspend and relax certain regulations to allow existing hospitals to rapidly increase space and capacity [19]. In effect, all New York's hospitals were ordered to put together plans to increase bed capacity by at least 50%, with a projected target to double overall bed capacity from 53,000 to 140,000. This order became one of the hospital's primary objectives to manage COVID patients. In New York, Government officials and hospital executives worked to expand finding space for more beds, deploying the Army Corps of Engineers to convert the Javits Center into a hospital and bringing up a Navy hospital ship, the USNS Comfort, to handle non-virus overflow cases.

The initial surge of 2020 demonstrated hospital unpreparedness for this mass influx of patients. The impact was felt across all patient care areas, including the Emergency Departments, in-patient beds, ambulatory beds, and all operations across the entire healthcare system. The pandemic required institutions to quickly expand and redefine the operational areas of hospitals expanding bed capacity and adding medical inventory with the expectation and requirement that the patient standard of care would not be impacted and regulatory standards maintained. This required that fundamental systems such as Nurse call, EMR integration, physiological monitoring, coupled with supporting communications, to provide data delivery, process acknowledgement, and documentation continued to be in place and operational in all expansion areas.

Specific impacts varied from hospital to hospital; however, all healthcare institutions were severely impacted and had to come to grips of dealing with the issues outlined below:

- Management of Large Patient Volumes
- Insufficient Patient Bed Capacity
- Insufficient Acute Care/ICU Level Beds
- Limited Patient Isolation Capability

- Shortages of Medical Device Equipment
  - Ventilators
  - Patient Monitoring
  - Air Filtering Systems
- Infection Control Issues
- Shortages of Staff and Patient Personal Protection
  - Personal Protective Equipment (PPE)
    - Face Masks and Shields
    - Gowns and Shields
- Shortages in Staffing and Availability due to
  - Contagion
  - Childcare Responsibilities
  - Transportation Limitations
- Shortages of General Supplies
  - Medical Supplies
  - Cleaning Supplies

Technology Driven Solutions

To attack these objectives, hospitals rapidly formulated collaborative inter-department working committees and groups to address the many issues facing the hospitals. In most cases, solutions utilized a diverse set of technology solutions, coordinated across all the many clinical areas of the hospital, requiring expertise from a broad range of clinical and technical staff. Hospitals attempted to expand and reconfigure the patient environment of care to maximize bed availability and provide the necessary medical devices to maintain the required standard of care.

At many hospitals, OR platforms were shut down for surgical procedures and converted into one of the critical care (ICU) expansion areas which effectively doubled the critical bed capacity. Similarly, other locations such as the Surgical Acute Care Unit (step down area) provided additional expansion and overflow capability which fortunately already supported many of the required clinical systems and medical devices such as physiological monitoring, telemetry, and advanced communications.

To support the increased bed/patient capacity, expanding medical equipment requirements were analyzed and correlated with the existing inventory to identify the specific equipment which needed to be acquired (purchase or rented). In today's Intelligent Hospital environment, medical devices are not simply standard-alone devices, but are complex medical systems that operate, communicate, and pass information via the hospital's network and wireless infrastructure, integrating with other medical devices and clinical applications. These complex medical data systems enable a seamless data environment, providing connectivity, data archival, and alarm delivery to the point of care, and ensure data integrity and association with the patient. Deploying medical equipment without this connectivity and associated integrations impacts both patient care and operational workflow. With all hospitals simultaneously attempting to obtain medical devices and supporting equipment, obtaining the appropriate devices supporting the necessary capabilities was challenging and frequently not possible. This required institutions to quickly alter, adjust, and update their patient care model and workflows, and utilize their technology groups to develop and establish temporary solutions and work-arounds.

To achieve these objectives, large quantities of medical devices and systems were redistributed, on-boarded, networked, and configured throughout the institution. It was critical that all devices were networked into the hospital infrastructure to maintain as seamless of data environment as possible. Since many hospitals in New York City utilize a common complement of medical equipment, this quickly resulted in shortages as outlined below.

Increasing Medical Equipment Inventory Requirements

- Ventilators
- Physiological Monitoring
  - Continuous
  - Vital Signs Intermittent
  - Telemetry
- Infusion Systems
- Feeding Pumps
- Air Quality Filtering Systems
  - Hepa filters

Management of Medical Equipment and Supporting Inventory Processes

During the pandemic, the acuity of patients rapidly changed between various patient care areas. It was critical that equipment be rapidly on-boarded, identified, acceptance tested, and device status determined and distributed as patients' distributions changed with the environment of care. With the shortages of medical equipment and the influx of a large quantity of devices and systems, which included new purchases, and temporary devices (such as loaners and rentals), the precise management of this inventory was vital to manage and support the continuing needs of the institution and patients. The hospitals centralized medical equipment database or archive needed to be accurately updated to provide detailed information of devices, availability, and location to all clinical areas. Even under pandemic conditions, each medical device needed to follow the regulatory standards, including acceptance testing and inspection by Biomedical Engineering prior to deployment. From a cybersecurity perspective, devices also needed to be assessed in terms of risk and vulnerability. All mobile and critical devices were also tagged with active radio frequency identification (RFID) technology to enable device real-time location and tracking (RTLS). The inventory needed to be globally visualized and available to all departments, staff, and users throughout the organization regardless of site or location.

The database enabled visualization via real-time location system (RTLS) technology which enhanced the management and distribution of medical equipment resulting from the fluctuations in COVID patient surges. Institutions which were RFID/RTLS enabled were able to capitalize on these enhanced inventory management processes. Memorial Sloan Kettering Cancer Center (MSKCC) has utilized active RFID-based systems operating on the hospital's robust Wi-Fi infrastructure for many years enabling the optimization of inventory management processes and workflow. Prior to the pandemic, the hospital's 60,000+ medical device inventory had 25% of the devices RFID enabled.

During the pandemic, the influx of critical devices, such as heap filters and ventilators, was required to be RFID tagged as part of the on-boarding process to ensure rentals, loaners, and government provided inventory were easily tracked, visible, and recoverable within the institution.

Minimizing Staff Exposure

During this period of patient overload with a high volume of patients requiring critical care, medical staff were also dealing with shortages of personal protective equipment and devices. With the closure of schools and transition to remote learning, childcare became a major issue for many younger parents and staff. In addition, public transportation systems were shut down, resulting in the lack of subway and bus transportation, limiting the ability for staff to travel back and forth to work. As the surge continued, many staff members also became infected or exposed resulting in the absence and quarantine of vital and critical staff. As the number of staff members unable to report to work increased, it quickly became clear that all clinical staff, especially clinical patient facing staff, needed to be protected as much as possible to minimize the growing staff shortage. As part of the patient care environment reconfiguring and expansion, several innovative steps were developed and implemented to minimize staff exposure by limiting the number of times staff needed to enter the patient room to review the patient's physiological parameters or handle equipment. To minimize this exposure, the following solutions were implemented within the patient care areas.

- Position Monitoring Systems Closer to Door Proximity
- Mount Monitoring Systems Outside the Room (Where Possible)
- Use of Remote Applications to Display Physiological Parameters outside the Room
- Install Extended Remotes
- Maximize the Use of Central Monitoring Stations and Slave Displays

These mobile and configurable solutions were deployed providing alarm and critical parameter delivery to the institutional EMR, data archives, as well as delivery to remote workstations (Figure 4.11) and handheld devices at the point of care.

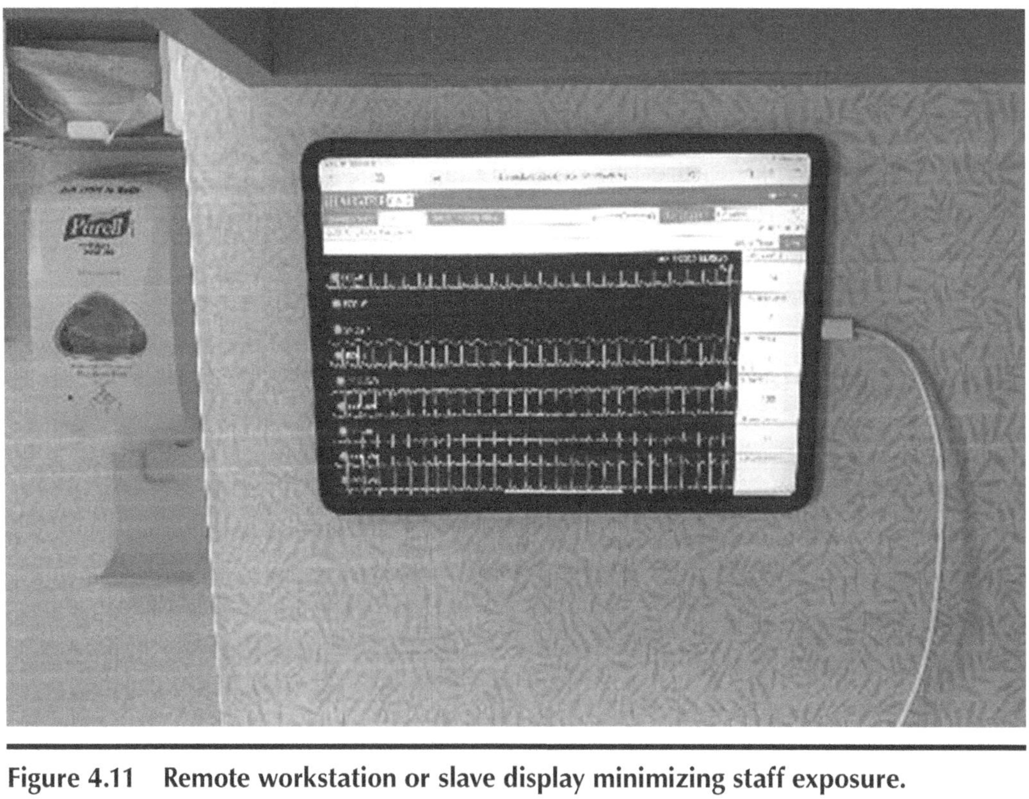

Figure 4.11 Remote workstation or slave display minimizing staff exposure.

Infection Control and Personal Protection Equipment/Devices

With high demand for personal protection equipment, dwindling supplies of this PPE continued to grow. MSK implemented a methodology of managing the distribution of PPE to staff while simultaneously investigating and preparing methods of reprocessing PPE and other devices. Biomedical Engineering investigated the combined use of UV and Hydrogen Peroxide Vapor sterilization (HPO) processes to reprocess various types of PPE. The Hospital worked with Central Processing Department (CPD), Nursing, Infection Control, and Environmental Safety to develop the process of managing the collection, association of PPE to specific staff, and redistributing the reprocessed PPE. In parallel, the hospital was designing and developing a dedicated reprocessing facility. The facility was designed and configured for both UV and HPO technologies and systems and was operationally set up to start a daily reprocessing. A significant workflow was designed for collection and redistribution of PPE maximizing staff safety and ensuring that all PPE was handled, reprocessed, and returned to the original users of the PPE. This facility was ready for operation as the shortages started to lift and PPE started to become increasingly available. The unit remains ready should future conditions warrant its use, with staff continuing to investigate and evaluate the effectiveness of the reprocessing methods.

The shortages also extended to supplies, where hospitals had limited access to disinfectants and hand sterilizing solutions. Novel approaches, which included internal preparation of hand sanitizer in our laboratories, ensured sufficient quantities of hand sanitizer for our patients and staff. MSK also deployed a novel personal hand sanitizer product worn by the clinical staff as illustrated in Figure 4.12. This product complimented our traditional use of sanitizers and enabled staff to enhance compliance with our infection control policies and standards.

Drivers for Change

Telemedicine

The COVID-19 pandemic highlighted the limitations and vulnerabilities of the standard healthcare model. Over the years, innovators have developed new tele-health models and practices. With the limitations being experienced

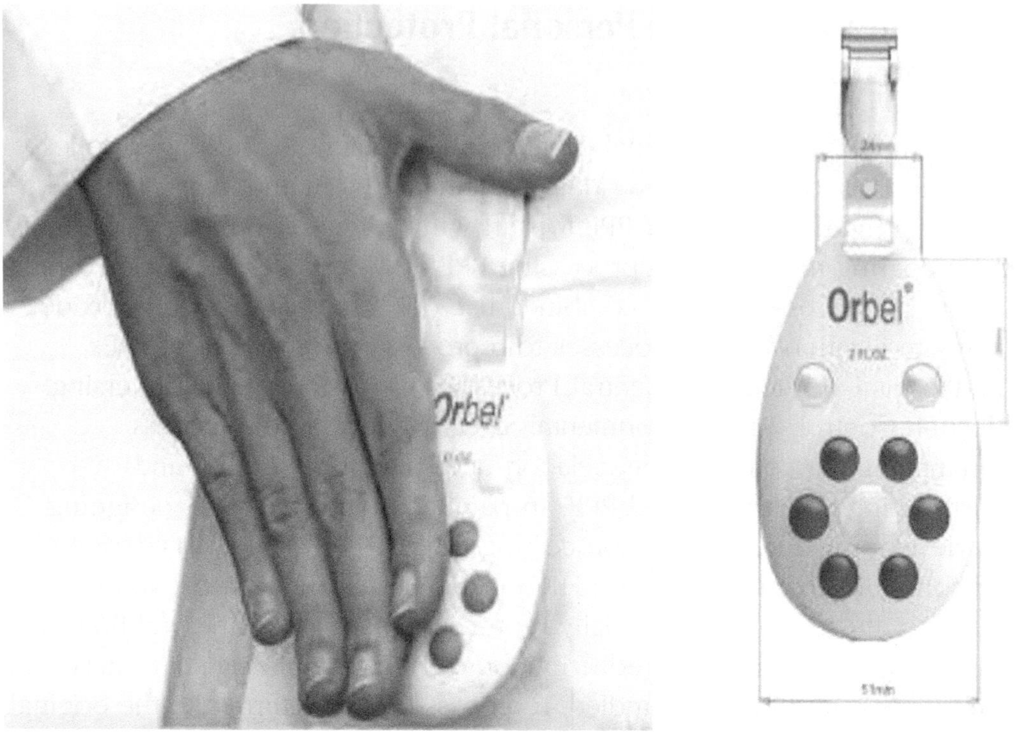

Figure 4.12 Point of care hand hygiene solution.

in the hospital systems and patient reluctance to come to large healthcare organizations, exploiting home health possibilities became a driver for treating patients remotely in their homes. The evolution of home-based networks and wireless capabilities like Wi-Fi and Bluetooth has enabled an assortment of medical devices to be used at home communicating back to hospitals over these networks. This enabled hospitals to manage patients remotely, limiting the number of patients requiring follow-up at the hospital, offsetting the patient load already being experienced in the Emergency departments and critical care units.

Data Analytics

The COVID-19 pandemic also heightened the importance and use of healthcare data and analytics in a way that has not been seen before. Accurate, reliable data and information played a critical role in providing healthcare executives and caregivers with the tools needed to make critical decisions. As Hospitals and health systems struggled to understand what might happen with COVID-19, hospitalizations continued to grow at an alarming rate. In-patient census in the United States for COVID-19 patients exploded in just 26 days from March 26 to April 21 according to statistics from the COVID Tracking Project [20]. Healthcare administrators scrambled to plan for unknown surges in hospitalizations, ICU admissions, and ventilator and other medical device needs. COVID-19 forced Healthcare systems to utilize real-time data coming from other countries as information from which to make key decisions. This data was the only reliable historical data that could be used to start projecting future cases, hospitalizations, etc. This data became the fundamental source for analytic prediction and decision support. This demonstrated the importance and value of analytic tools utilizing real-time data sources to drive administrative decisions and clinical care solutions.

COVID-19 Driven Engineering Development

In traditional hospitals, Biomedical Engineering's role is primarily focused on Health Technology Management or Clinical Engineering emphasizing medical equipment management, service support, maintenance, and preventative maintenance ensuring regulatory compliance and addressing the environment of care initiatives of CMS and the Joint Commission. As was previously discussed, this expertise played a significant role in equipment expansion and management during the pandemic. However, MSK Biomedical Engineering, with a strong focus on the investigation of new and evolving technologies, was able to exploit these resources to impact patient care, enhance outcomes, and ensure improved patient safety and satisfaction.

At MSK, Biomedical Engineering, supporting strong synergies and affiliations with IT and clinical departments, focused its resources to provide analytic, design, and development capabilities exploiting and implementing new technologies and innovations. Biomedical Engineering was augmented to include electromechanical and software application development and production capabilities which included Mechanical Engineering and Biomedical Systems groups providing capabilities in mechanical design, CAD, manufacture, basic research, electrical and electromechanical design project management, and software applications development. These groups operate the Institution's dynamic simulation laboratory (SimLab) as well as the clinical 3D printing laboratory adding a new dimension to exploiting technology and engineering development. The SimLab provides a unique dynamic design and development testbed enabling clinical device and software upgrade testing, and the evaluation of new devices and interfaces, cybersecurity vulnerabilities, new technologies, the creation of new workflows, and staff training in an isolated environment minimizing patient safety and risk. The 3D printing laboratory, originally implemented as a rapid prototyping and production tool, was quickly focused and enabled patient specific treatment solutions, such as anatomically accurate models, custom surgical guides, and tooling.

Responding to a Crisis with 3D Printing, Design, Engineering, and Production

With the onset of COVID-19 pandemic, the value of this type of in-house capability and functionality was quickly realized and demonstrated. Innovations both of our own initiation and those defined by groups within MSKCC and outside organizations were quickly designed, prototyped, evaluated, and developed for production. Dealing with uncertainty and supply chain failures, these teams designed and implemented medical devices, PPE, and novel solutions to meet the needs of the hospital staff and patients. COVID-19 demonstrated the need for immediate parts, mechanical devices, and medical devices.

As the COVID-19 pandemic exploded within the hospital, these biomedical investigation and development groups were targeted to support the most critical needs of our hospital. This capability quickly focused on the development of COVID-19-related innovations which included the design of medical devices, PPE, and novel solutions to meet the escalating needs of the hospital staff and patients. The following projects were successfully implemented and deployed to aid in the battle against COVID.

- Modification of Snorkel Masks for Use as PPE
- Filter Adapters for 3M Face Masks
- 3D Printed Face Shields
- 3D Printed Testing Swabs
- HEPA Filter Conversion to Spot Ventilation
- MSK Octopus: Multi-Patient Ventilator Splitting
- Crisis Vent: Emergency-Use Ventilator
- Intubation/Extubation Aerosol Boxes

Ventilation Support Systems

As previously discussed in the early stages of the pandemic, hospital systems were overwhelmed with patients needing mechanical ventilation and a consequent shortage of ventilation systems [21]. A number of solutions were proposed to meet the demand: emergency production of new ventilators (both by existing manufacturers and non-medical/automotive manufacturers), home-made ventilators created by medical providers or universities, alternative ventilation solutions, and splitting/sharing of existing ventilators. An immediate solution enabled rapid turnaround focused on the splitting of ventilator systems between patients. In March of 2020, when New York State approved emergency use of ventilator splitting for COVID-19 patients [22], a collaborative MSK team with members from Biomedical Engineering, Anesthesiology, and Surgery came together to work on this challenge and established the requirements. The proposed solution, shown in Figure 4.13, incorporated the use of an adjustable, fixed-pressure regulator on each parallel ventilation circuit and allowed completely independent adjustment of inspiratory pressure for every patient [23].

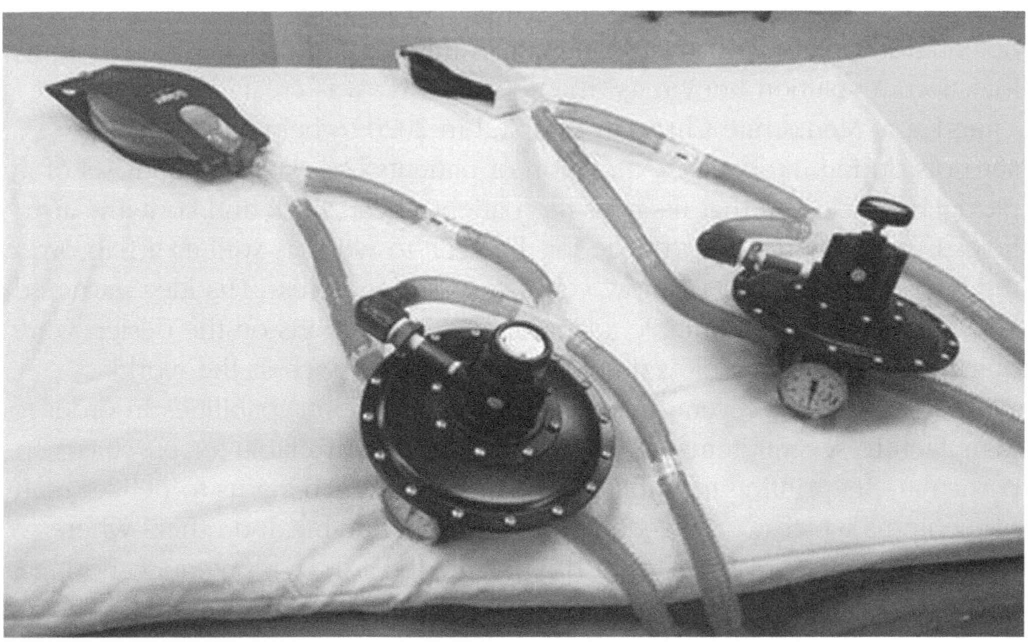

Figure 4.13 Ventilation splitter – enabling multiple patients/ventilator.

Crisis Ventilator

In a parallel effort, a team consisting of Surgeons, Biomedical Engineers, Respiratory Therapists, Anesthesiologists, and Device Development Engineers came together to design and build a prototype of what we characterize as an emergency-use crisis ventilator or "Crisis Vent." The intent of this device was to provide a compromise between functionality, simplicity, and ease of building to facilitate the broadest possible distribution of lifesaving care in the event of a traditional equipment shortage. This was not a traditional ventilator, but rather a method to provide a bare minimum of functionality to maintain a sick patient when no other option exists.

3D Printing and PPE Development

Intubation/Extubation Boxes

Various types of PPE were promptly adopted or developed to protect patients and healthcare providers (HCPs) from airborne viral exposure to COVID-19. Clinicians working with a patient's airway were objectively determined to be in the highest risk category. One area of particularly high potential exposure risk for HCPs is during intubation/extubation of patients with COVID-19. The insertion and, even more so, removal of endotracheal tubes can generate aerosols exiting the patient's mouth and are thought to potentially contain high viral loads which might put those nearby at risk of infection. A solution known as "the aerosol box" was created by Dr Hsein Yung Lai at Mennonite Christian Hospital in 2020 to help contain these aerosols during intubation/extubation of patients [24]. The box consists of a plexiglass enclosure that fits over the patient's head/neck and contains arm holes in the superior (head) side for the HCP to work as well as a flap on the inferior (chest) side to allow ventilation tubing access. His idea garnered a lot of attention and multiple adaptations and iterations on the design were quickly created and deployed at MSK and by teams across the world.

MSK was able to leverage our in-house fabrication capabilities in order to manufacture several items which were not readily available for purchase. In particular, 3D printing technology was critical in enabling us to deliver many of the items which were needed early in the pandemic and others where we were able to proactively plan for the heightened risk of a potential shortage in the future. Our Biomedical Engineering group used 3D printing in the production of face shields, mask adapters, respirator adapters, ventilation tubing adapters, and nasopharyngeal test swabs, among others. More than

700 face shields were internally produced over the course of a month, 300 more were printed and donated to us by non-medical business partner companies, and this kept many of our healthcare workers protected until we were again able to purchase them from commercial suppliers.

Case Study: 3D Printed Nasopharyngeal Test Swabs

Another critical supply which was difficult to source in early 2020 were nasopharyngeal test swabs used for COVID-19 testing. A collaborative effort between a printer manufacturer (Formlabs) and two hospital groups (Northwell Health and USF Health) yielded a design which could easily be printed and sterilized for clinical use (Figure 4.14) [25]. We were able to quickly set up a system to print and process up to 2,000 swabs daily. Within a few days, we made test prints, ran through sterilization, and had our internal teams validate efficacy and approve for use.

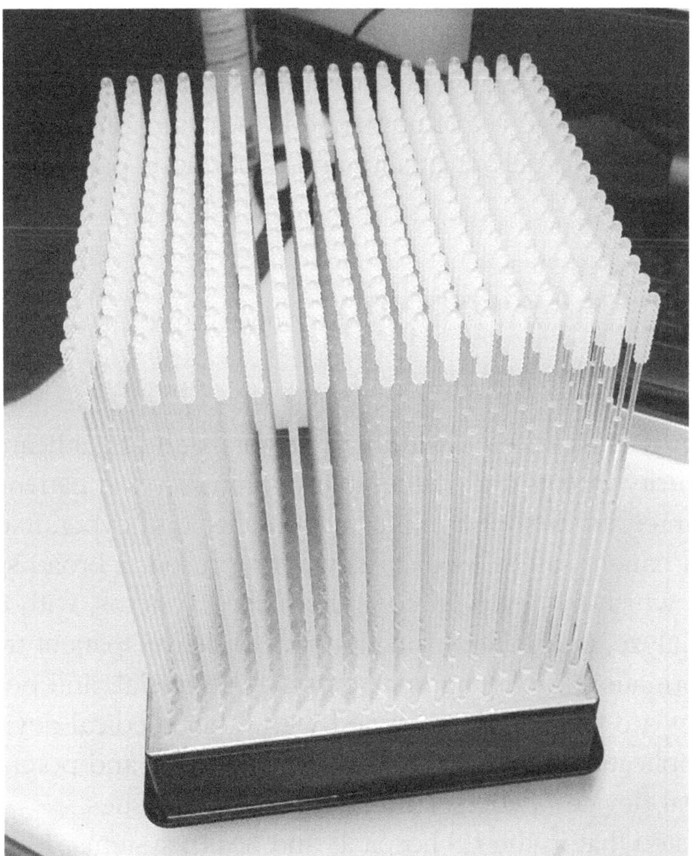

Figure 4.14 3D printed nasopharyngeal test swabs.

Emergency Preparedness: Lessons Learned

The events of the Coronavirus pandemic have clearly identified the need for expanded emergency preparedness and preparation for the unexpected. These events can be natural events, such as hurricanes, floods, or earthquakes, or deliberate events, such as terrorist attacks. As we read through the chapters and reflect on the steps that were taken by the many institutions across the United States and the world, it becomes clear that the successful implementation of solutions is largely dependent on having specific technologies, capabilities, workflows, and plans in place prior to any event occurring. Though the world has moved to supply on demand business model, COVID has demonstrated that there is a realistic need to maintain at least a minimum level of supplies or maintain in-house capability to design and produce alternative PPE, hygiene products, etc. to support unplanned events. Fortunately, the pandemic did not affect power and utility disruptions; however, it is easy to envision how these could have significantly further impacted our COVID response and further introduced risk to patient outcomes and safety. Any emergency preparation planning needs to address a broad set of unplanned events enabling healthcare systems to continue to deliver quality and continuous patient care.

Medical Device Cybersecurity

Introduction

One of the growing concerns within all hospitals and healthcare systems are cybersecurity threats and vulnerability to impacts on patient care, operational integrity disruption, information loss, and cyberattack resulting in impact on patient safety. Medical devices represent a broad spectrum of technologies which has evolved computer-based systems, with capabilities to collect, analyze, and control functionality to deliver patient treatment and transmit information, including demographics, financial, and personal health. These devices are frequently integrated with other medical devices as well as clinical applications, further broadening the scope and potential severity of the medical device's cybersecurity concerns and issues.

It is estimated that in the US hospitals and health systems, 10–15 million medical devices are deployed and operational, with 25–50% of these devices networked either hard wire or wirelessly. Examining the magnitude or scope of medical device security threats in today's typical academic Intelligent Hospital for every 25,000 medical devices used in the patient care applications results in a minimum of 2,500 devices having vulnerability to penetration or cyberattack. Within the Intelligent Hospital, the complex infrastructure and medical systems represent two primary scenarios of concern requiring specific processes to be implemented to minimize vulnerability and risk. First is an attack directed at specific medical devices or systems focused on disrupting device operation and functionality. This clearly has a direct impact on patient safety and has the potential of shutting down specific and critical operations. This can include treatment devices, such as linear accelerators, which directly impact radiation dose putting patients at risk, or infusion systems, medication delivery, diagnostic devices providing corrupted or incorrect decision support data or information, and care/monitoring devices, such as physiological monitoring. For critical devices used for life support, such as ventilators, anesthesia units, or pacemakers, failure can result in severe injury or death. These types of attacks are focused to disable or directly impact clinical operations triggering a potential ransomware attack. The second type of cybersecurity concern is that these complex systems include integrations and interconnected systems providing many points of access to the hospital's network enabling access to the broader healthcare systems infrastructure. These cyberattacks target and can attack a broad base of hospitals' applications and information systems, clinical, business, and financial systems. Hospitals deploy and operate tens of thousands of medical systems throughout the hospital enabling thousands of points of access with minimal or limited security.

This vulnerability to cybersecurity intrusion or attack has placed new demands on the management of network infrastructure and the medical systems/devices running on the network, to ensure these complex systems are assessed to identify vulnerabilities and upgraded with specific controls to minimize the potential of these types of attacks. This is a complex and tedious process typically requiring a significant amount of planning, budget, staffing, and a robust workflow to successfully implement and manage.

Medical Device Cybersecurity Management Process

In today's Intelligent Hospital, all business and clinical operations must include a dynamic, robust, and diligent medical device cybersecurity risk management program to ensure operational integrity, business continuity, and patient security and safety. The objective of this program is to define and establish operational policies, guidelines, and procedures to effectively manage, mitigate, and minimize risk. This requires adaptive long-range planning and administrative support, funding, and well-defined and established day-to-day processes to monitor, continuously mitigate new and evolving risks, vulnerabilities or threats, and respond to active threats ensuring recovery and operational integrity and continuity.

The process of continuously reviewing and managing medical devices' operation status to ensure data quality, patient and information safety, and operational integrity is critical to an institution's business process and requires a broad base of stakeholders to determine the specific direction and objectives for analyzing, determining, and resolving threat vulnerabilities. Ideally, the hospital defines and implements the processes used to manage threats and vulnerabilities through a Healthcare Technology Risk Management Committee consisting of many stakeholders and participants from the many departments across the institution.

The fundamental process can be generally outlined by the following tasks and associated workflows to analyze and evaluate all devices in terms of device type, functionality, vulnerability, and risk potential utilized in the institution. The process includes the following steps:

- Risk Assessment
  - Process to collect information about the status of devices, establishing qualitative or quantitative vulnerability probability and rating.
- Risk Identification and Vulnerability Management
  - Identify processes to monitor real-time device status including ongoing vulnerabilities and potential threats.
- Risk Mitigation
  - Identify the specific vulnerabilities and define measures or controls to minimize the impact of these vulnerabilities to potential threats and attacks.
- Response Procedures and Recovery
  - Establish procedures to manage vulnerabilities and respond to active threats to ensure operational integrity and recovery.

The effectiveness of these processes requires ensuring that all systems are assessed and managed to minimize vulnerability, which requires a robust multidisciplinary approach and team focused on identifying and managing life cycle providing a continuous review analysis and review of changing threats and a process ensuring the implementation of upgrades, controls, and a real-time and continuous monitoring of all components of the hospital's digital environment as outlined below.

- Infrastructure (Wired and Wireless)
- Medical Devices
- Clinical and Supporting Software Applications
- Integrations and Supporting IoT Devices

The specific tasks or functions associated with managing the vulnerability and security risk associated with medical systems and supporting assets include:

- Ensure a Dynamic and Accurate Inventory Management Program
  - Unique Identification
  - Location Tracking
  - Ownership
  - Comprehensive System Functional Descriptions
  - Historical Documentation
- Provide an Ongoing Inventory Hygiene Program
  - On-boarding or Acceptance Testing
  - Assessment Processes
  - Device Decommissioning
  - End of Life Management
  - Upgrades and Patches
  - Controls Implementations
- Support Real-Time and Dynamic Monitoring of Medical Systems for Vulnerabilities and Potential Threats
  - Vulnerability Identifications
    Threat Identification and Alert
  - Missing Controls
  - Out-of-Date Software
- Develop an Advanced Testing and Simulation Process
  - Test and Evaluate Control Effectiveness
  - Development of Response Procedures
  - Establish Recovery Protocols and Processes

Proactive Institutional Medical Device Cybersecurity Risk Management Program

The management process begins with an organizational commitment, structure, and effective governance to establish and ensure an enhanced process for establishing an accurate inventory which is validated at regular intervals within the hospital and supporting clinical areas, in-patient, ambulatory care, clinical laboratories, treatment areas, and supporting locations. The inventory management process includes the following:

- Medical Device Specification and Selection
- Medical Device On-boarding
  - Unique Identification and Classification
  - Incoming Acceptance testing
- Assessment
- Life Cycle Management
  - Upgrades and Recall Management
- Replacement Planning
- Operational Functionality
  - Network Connectivity (Wired, Wireless, Proprietary)
  - Integrations
  - Data Flow
- Medical Device Tracking
- Asset/Medical Device Repository or Database Archive

Fundamental is the documentation associated with the inventory management process and medical devices. In the Intelligent Hospital, the most referenced archive is the medical system or equipment database used to support the Biomedical Engineering operations, focused on ensuring medical device operational integrity and regulatory compliance. Using the Biomedical medical database (CMMS archive is an ideal place for defining and documenting any potential patient safety risks and cyber vulnerabilities. The starting point of any security operation is focused on identifying all the assets, medical systems, component assets, and information associated with the cybersecurity program. Maintaining an accurate and dynamic inventory ensures a robust and ongoing risk assessment process fundamental to the cybersecurity program. The asset or inventory program requires an enhanced on-boarding and decommissioning process, which includes a security assessment process which is directly associated with and part of the purchase decision. This inventory management process must be sufficiently robust to ensure all incoming systems, devices, and components undergo a cybersecurity assessment as part of the traditional incoming evaluation and acceptance testing process prior to deployment into clinical operation and patient care use.

The inventory needs to define the overall system hierarchy as illustrated in the system block diagram (Figure 4.15), identifying a system level view which includes multiple subsystems, the component medical devices, the required infrastructure, servers, interfaced or integrated clinical applications, and supporting IoT devices. This system level block diagram defines all the pathways through which a security penetration or attack can enter and pass between systems over the infrastructure. The block diagram identifies and defines how many individual component assessments are required for the overall complex medical system.

As illustrated in the physiological monitor system, the system consists of two component monitoring platforms, several models of hardwired physiological monitors, a telemetry subsystem, central stations, multiple gateways, servers, multiple integrated clinical applications, ECG archival, connectivity solution, remote monitoring application, electronic medical record, etc.

In this case, the ability to fully visualize and assess the physiological monitoring platform/system for its overall vulnerability would require the following analysis, evaluations, and assessments:

- Physiological Monitor Platform 1 Acute Care/ICU
- Physiological Monitor Platform 2 General Patient Monitoring

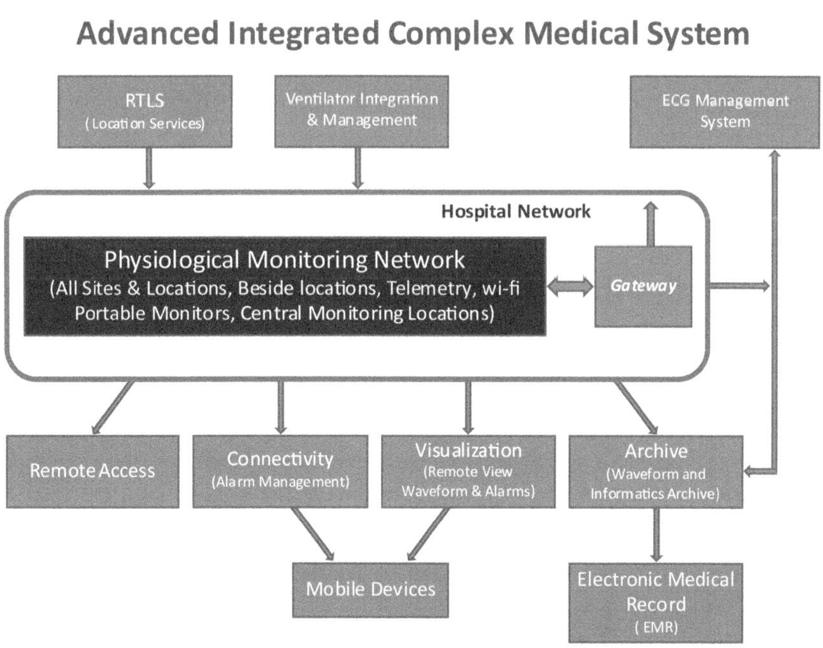

Figure 4.15 Advanced complex medical system hierarchy.

- Telemetry Unit
- Central Station
- Gateway
- Servers
- Clinical Applications
  - ECG Archival Systems
  - Connectivity Solution
  - Remote Monitoring Dashboard
  - Electronic Medical Record

Note: These clinical applications are also likely to be integrated or interfaced with other medical devices/Systems operational in the institution, which would then impact or redefine the assessment process required by these clinical applications.

Inventory Hygiene

To minimize current and future vulnerability, the Intelligent Hospital needs to support an ongoing hygiene program which includes review and updates to patching software, updating hardware, version control, and a clear definition of controls to minimize vulnerability. Similar to a medical device preventative maintenance program, a cyber review process requires periodic review and reassessment based on the changes to the system to identify and evaluate the current state of the devices. The issue of implementing upgrades, patches, and controls can directly introduce and require budgets and other financial issues. Biomedical Engineering or other managing body needs to continuously monitor and identify controls and costs and prioritize implementation based on available funds, end of life replacement plans, etc. These processes need to be well defined, documented in the medical device database and supporting service history.

Real-Time Monitoring

Any robust medical device cybersecurity program must include a comprehensive process to dynamically monitor network activity, identify all devices operating on the network, and scan devices and the network for vulnerabilities. The monitoring process periodically analyzes the network data traffic to determine if there are any potential risks to the environment. The vulnerability scan identifies any weakness that increases device vulnerability, including the computer operating system, installed software, current patches, exposed ports, and applications to determine weaknesses such as lacking controls or potential threat activity.

To identify and minimize cyber vulnerability, potential threats, define controls to enhance device hygiene, and maximize rapid response, the Intelligent Hospital will need to incorporate and support an advanced real-time network monitoring system. These systems examine medical device configurations, settings, operational characteristics, and data traffic of all medical devices and supporting clinical applications operating on the hospital network and infrastructure to identify vulnerabilities and or/validate operational integrity. This implementation may include multiple monitoring applications that will continuously scan and assess the infrastructure, servers, firewalls, devices, and systems for vulnerabilities, including missing controls, updates, recall or end of life, as well as for potential intrusions and attacks. These systems will generate near real-time alerts and alarms resulting from unexpected activity. Unique response procedures will need to be developed as a function of the type of device and the nature of the alert to minimize the impact of penetrations and attacks.

Risk Assessment Process

The assessment process functions to define and establish a management protocol with a scoring criterion, establishing a prioritization methodology which can be based on many factors to define a subjective device risk level. This scoring model is typically based on the following factors:

- Inventory Management Accuracy
- Device Functional Description and Informatics (PHI, Sensitive or Critical Data)
- Device State (State of the Art, Legacy, or End of Life)
- Device Capabilities and Functionality
- Ecosystem (Security Level of the Physical Area)
- Physical Access
- Vulnerability Control Measures in Place
- Operation Risk

The assessment process requires a diligent review of the device functionality and capabilities, access, data ports, and communication protocols. Consequently, the assessment has identified many new parameters which can and should be used to assess or score the threat vulnerability. A robust program would require that these parameters be associated and documented within the medical device database and routinely monitored and updated. (The database inventory parameters detail the information associated with medical systems and medical device management as identified in Appendix 2).

The cybersecurity assessment process requires a significant increase in the detail and amount of information required about the devices to assess device vulnerability, as summarized below.

Manufacturer/Vendor

- MDS details Devices' Cybersecurity Feature
- Manufacture Residual Risk File
- Operating Software/Firmware
- Integrated or Component Commercial Off-the-Shelf Software
- Required Security Software (Anti-Malware, Intrusion Protections, etc.)
- PHI
- Other types of Software (Images, Physiological Waveforms, etc.)
- Encryption (Stored or Transmitted)
- Malware or Code Obfuscation Used
- Authentication Used
- Device Network Communications Properties
  - MAC, IP Addresses
  - Network Configuration and Protocols
  - Connectivity
  - Device to Device Connectivity
  - Network Type (VPN, Wired/Wireless, Bluetooth)
- Access (Removable Media, Port Definitions)

Ecosystem/Operational Environment

- Location Security (Security Location, Locked Areas)
- Network Architecture (Isolated, etc.)
- Remote Access (Users/Vendors)

Device Management

- Policies and Procedures
- Access Management
- Encryption
- Change Management
- Life Cycle Management
- Selection/Purchase Management and Control (Standardization)

- On-boarding/Decommissioning
- Preventative Management
- Recall Management
- Education and Training

Risk Level

Risk level is commonly defined as a function of or probability of occurrence or incidence and the magnitude or severity level. The determination of the "Risk Level" is based on a combination of assessment results. The first is risk in terms of severity to the patient safety and patient outcome. The other pathway introduces operational integrity, financial risk, institutional reputation and credibility, and potential legal consequences.

The assessment process defines the contributing factors which directly impact the determination or likelihood of occurrence, the severity or impact to the medical systems, impact to patient care and business operations and patient safety. These are also directly related and determined as a function of the device type, functionality, extent of the vulnerability, and potential impact to the hospital as outlined:

- Device Type
  - Life Support
  - General Patient Care
- Functionality
  - Diagnostics
  - Treatment Delivery
  - General Patient Care
- Quantity and Distribution
  - Institutional Standard Applicable to all Patient Care Areas
- Type of Data Compromise
  - Operation Integrity
  - Data Availability
  - Data Integrity
  - PHI
- Potential Patient Impact or Severity of Harm
  - Catastrophic
  - Critical to Operations
  - Marginal
  - Negligible

Factors Impacting Risk

Risk is further defined as inherent risk, the risk of the medical devices operating on the hospital infrastructure, and residual risk resulting post implementation of specific controls which mitigate potential risk concerns. The ability to mitigate risk becomes a function of an institution's effectively implementing controls, and becomes a function of budget, functional capability, medical devices age, availability of controls, and overall prioritization typically defined in terms of capital spending.

Severity Scoring Criteria

Prioritizing an institution's cybersecurity mitigation is established and defined based on the scoring criteria, and the unique institution's specific weighted computations which aid the hospital to make critical decisions and prioritize remediation which is based on an institution's acceptable "Risk" level. This determination of acceptable risk level is commonly based on the inherent vulnerability, probability of occurrence, severity, and impact to patient safety (defined by Grimms [26]) as a function of increasing probability of occurrence vs. increasing severity.

Ideally, during the medical device technical review process prior to purchase or in the worst case during the incoming acceptance testing and inventory process, all medical systems or devices are identified in terms of networked or stand-alone operation determining vulnerability. Systems classified as networked or network capable require a dynamic assessment to identify and define device vulnerabilities categorized as a function of risk level. New institutional or updated policies must include advanced procedures and response protocols defining how Biomedical Engineering, IT, and user department will coordinate the response to managing suspicious activity and potential attacks as well as provide systems operational attack and response recovery.

Response and Recovery

The process of responding to cybersecurity threats focuses on several primary actions

1. Identification of threat activity
2. The specific response to threat activity
3. Recovery of System Operational Integrity

Advanced Testing and Simulation Laboratory

Within today's Intelligent Hospital, medical systems/devices are increasingly complex coupled with layers of inter-connectivity, integrated communications networked with and supporting wireless operation creating an operational inter-dependence. Within this dynamic patient care environment, modifications in hardware and/or software, including upgrades or patches to medical devices, application software, and commercial devices, have the potential to negatively impact the operation or functionality of any of the system components or the entire system. In addition, the inclusion of new devices, software, commercial devices, and other supporting technologies can directly impact operational integrity. The consequences of this unforeseen interaction can negatively affect data integrity, accuracy, and consistency, which has the potential to inadvertently affect other systems or applications directly impacting patient safety.

In the past, devices were managed, supported, and tested directly within the environment care. However, today's highly integrated Intelligent Hospital environment dictates the need for new strategies for support to be considered and developed. It becomes critical to be able to service, make upgrades, develop enhanced procedures, and evaluate new devices, enabling technologies and clinical applications in an isolated environment prior to deployment institutionally within the patient care environment (Figure 4.16).

Ideally, the Simulation Laboratory (SimLab) duplicates the patient environment of care, in terms of network infrastructure, medical devices, clinical applications, connectivity solution, IoT or handheld devices (Wi-Fi phones, iPads, etc.), and RFID/RFID/BLE, and functionally represents the patient care environment. The simulation facility will require multiple types of patient care and support environments to simulate overall Intelligent Hospital technologies, workflows, and operations as illustrated in Figure 4.17. Critical

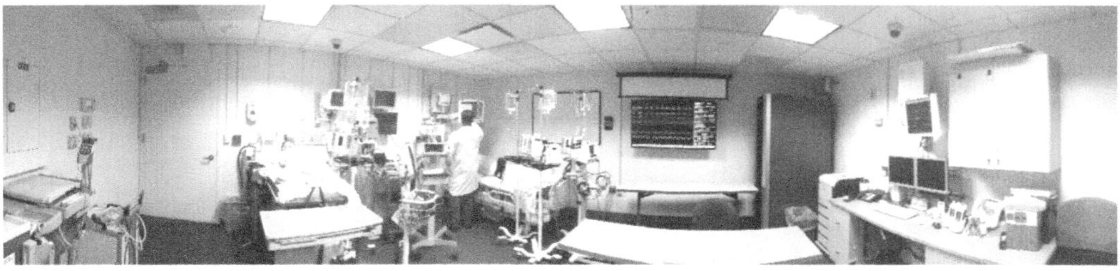

Figure 4.16 Intelligent Hospital Simulation Laboratory (SimLab).

to the operation and functions that will occur in the simulation laboratory, it is critical that the simulation be able to maintain normal operation as part of the hospital's infrastructure and be switched to an isolated environment to perform testing including new upgrades, hardware and software, and dynamic cybersecurity simulation. The SimLab enables expanded testing and maintenance capabilities and procedures to be developed presenting a new dynamic and comprehensive tool to evaluate various hardware configurations and options as well as evaluating workflow to manage these complex systems. The SimLab provides the resources to critically test and validate processes and changes to ensure overall system operational integrity and functionality. This Simulation Laboratory is ideally configurable to enable various patient care environments (OR, ICU, Radiation Oncology, etc.), as well as support areas, to be dynamically simulated as needed by the hospital. This dynamic test environment also enables a broad set of tests and evaluations to be performed

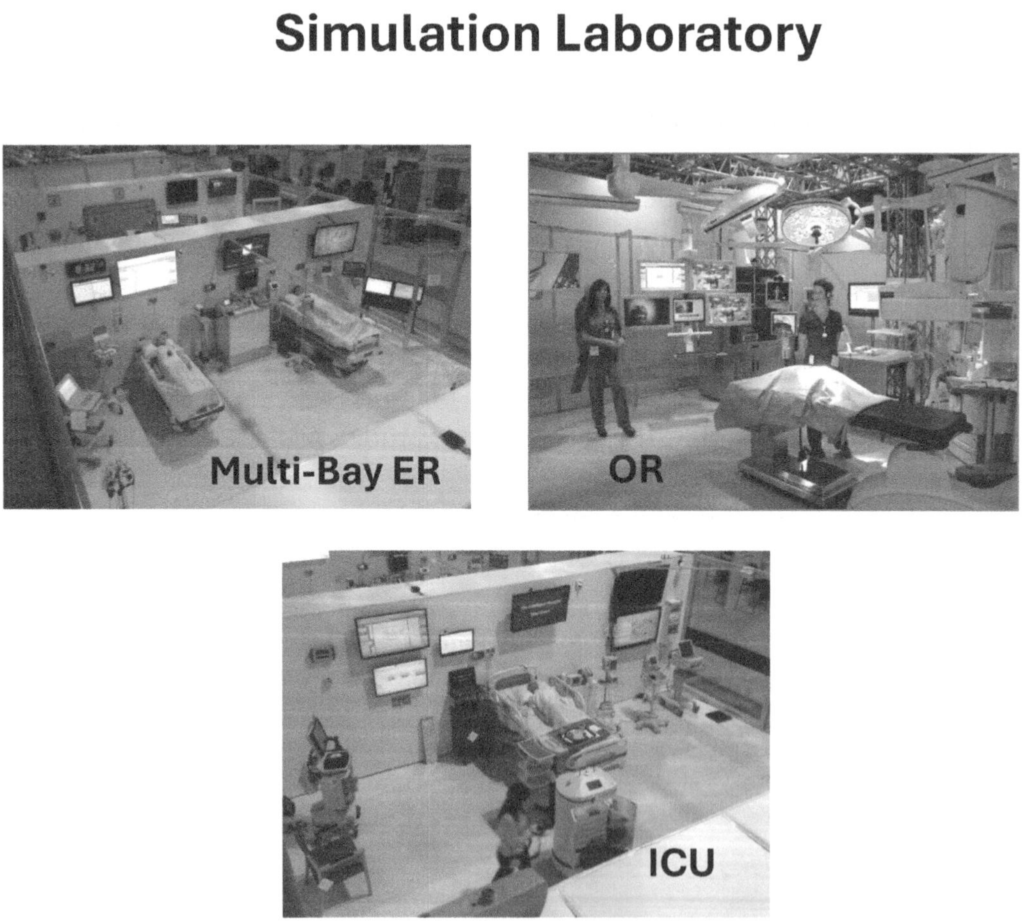

Figure 4.17 Patient care environment simulation laboratory.

including integration of new medical devices, applications, and enabling technologies, such as BLE, to develop and evaluate new use cases and workflows. The SimLab uniquely provides capabilities to perform research into the continuously evolving cybersecurity threats to understand vulnerability based on how cyberattacks penetrate hospital systems and infrastructure. The lab enables controls to be implemented and evaluated minimizing vulnerability and enabling dynamic response and recovery focused on the development of threat response in a safe and contained environment.

Reengineering Health Technology Management and Support Advanced Biomedical Engineering

Introduction

The rise of the Intelligent Hospital has evolved through a continuous evolution and development of advanced technologies including medical devices, clinical applications, hospital infrastructure supported through large-scale integrations between these devices, clinical applications, and exploiting a variety of enabling and diverse IoT technologies. These solutions focus and enable the creation of a seamless patient care data environment establishing a continuum of care, enabling the rapid exchange of patient specific, validated, and actionable information. This dynamic environment provides the ability to develop and deploy state-of-the-art complex systems which directly and positively improve patient outcomes, enhance patient safety, and positively impact the patient experience and satisfaction. Dynamic institutions maintain their edge by being at the forefront of research, innovation, and technological advances, rapidly transitioning these diverse areas into diagnostics, treatment, and the delivery of patient care across all aspects of the patient care environment. This continuous evolution defines the fundamentals for the continued evolution and development of the Intelligent Hospital.

This continuous evolution and integration of new technologies is frequently not equally considered, overlooked, or planned as part of the management and support of the Intelligent Hospital. It is critical that the planning for the Intelligent Hospital include a broad base of technology research, investigation, and, equally important, the management and support solutions and modalities providing a robust maintenance and management strategy incorporating the necessary broad spectrum of disciplines and expertise. A critical and fundamental part of any Intelligent Hospital initiative must include a comprehensive plan to establish and reengineer an adaptive Biomedical Technology development and management group incorporating a broad base of technology and engineering expertise coupled with operational synergies and collaborations at many levels within the hospital. To successfully deploy these solutions to optimize and customize a mix of technologies requires Biomedical Engineering to transform from its traditional Clinical Engineering role focused on the repair, preventative maintenance, medical equipment management, and regulatory compliance. The broader scope of Biomedical Engineering in the Intelligent Hospital needs to include the design, development, and advanced multidisciplinary engineering functions, including mechanical/electromechanical, electrical engineering, computer science, and software development, as well as supporting expertise and knowledge in materials science, physiology, anatomy, and computational analytics. To achieve this goal, Biomedical Engineering must be reengineered transforming its operational objectives and expanding its functionality to become a driver and direct and effective contributor to the overall healthcare process.

Definition of Biomedical Engineering

Within the Intelligent Hospital, Biomedical Engineering includes a mix of traditional engineering skills coupled with biological sciences, information management, and computer sciences targeting advanced patient specific and centric medical solutions which can be broadly defined as follows:

> Biomedical Engineering is the application of engineering and computer science principles to solve medical and biological issues and problems.

or, more applicable in this case,

> Biomedical engineering is a discipline that advances knowledge in engineering, biology and medicine, and improves human health through cross-disciplinary activities that integrate the engineering sciences with biomedical sciences and clinical practice.

Biomedical Engineering functions and focuses on the design, development, and implementation of solutions which are commonly subdivided into several specific phases or areas as outlined below:

- **Research:** This refers to the study, investigation, and development of or understanding of the theoretical scientific or engineering principles.

- **Systems Design:** Higher level functional design; commonly identifies the set of problems and associated disciplines needed to resolve specific design problems. The system design phases can include multiple engineering and scientific disciplines, including mechanical, electrical, chemical engineering, materials science, computer science, and a variety of analytics where the solution interfaces with human interaction for control and visualization.
- **Design and Development**: Focuses on transitioning the conceptual systems designs to detailed integrated design defining the implementation and development of components and fabricated or implemented solution, system, or clinical application.
  - **Electrical Design:** Design and development of all electronic systems (Analog, digital, RF).
  - **Mechanical Design:** Design of mechanical systems used in patient care and patient specific solutions as described in detail in Section 3.
  - **Software Design:** Development of the software algorithms to acquire (collect), convert, process, store (archive), and display information.
- **Engineering Management:** Provide the technical and business tools to monitor, control, and manage medical/clinical solutions and patients.
  - **Business Management:** Investigate and control the cost of materials, research, design, development, and manufacture, based on the projected budgets developed and the projected cost of marketing and sales.
  - **Production/Manufacturing**: Focuses on developing the techniques and processes to produce a product. These products can be electric devices, biologic products, such as pharmaceuticals, foods, chemicals, etc.
  - **Support Engineering:** All industries and business require engineering, communications, and computer science functions to use to maximize deployment. However, support engineering can best be defined as engineering functions outside the primary business line of the company. The examples we will be primarily interested in will focus on the medical industry.

Biomedical Engineering Disciplines

Biomedical Engineering is a highly interdisciplinary science, closely integrating engineering with the applied sciences of medicine, biology, chemistry, and physics and the formal sciences like mathematics, computer science, and social sciences. Biomedical Engineering encompasses theoretical research, investigation, and development of new and evolving technologies, simulation, modeling, design, development, fabrication, and implementation of specific solutions. Within the medical or healthcare environment, Biomedical Engineering is focused on many distinct areas of interest and defined to include functionality and expertise in the following areas:

1. **Biomaterials**: The study, design, and development of implantable and replacement materials.
2. **Biosensors**: Sensors designed to detect biologic events including the conversion of these events to biological signals.
3. **Instrumentation and Medical Device Design**: The design and development of sensors and electronics to monitor and measure physiologic events, such as ECG, blood pressure, temperature, etc.
4. **Prosthetic Devices and Artificial Organs**: The design and development of devices and systems for the augmentation and replacement of bodily functions such as artificial heart and limbs.
5. **Imaging**: The development of graphical visual displays detailing human anatomy and physiologic functions.
6. **Biotechnology:**
   - Development of Antibiotics
   - Genetic Engineering
7. **Rehabilitation Engineering**: Design and development of devices, systems, and procedures for therapeutic and rehabilitation purposes.
8. **Medical Informatics**: Management of patient-related data, data association and interpreting results, and providing decision support tools to assist in the clinical decision process. These systems range from database/archive systems to expert systems, neural networks, and artificial intelligence.
   - Clinical Information Systems
   - Networks, Data Transfer Protocols
   - 3D Imaging and Reconstruction
   - 3D Modeling and AR/XR Visualization
9. **Biological Effects:** The study of the effects and consequences of electrical and magnetic fields on biological tissues and their functions.
   a. **Biomechanics**: Study of static and fluid mechanics
      - Study of head and neck response to transitory acceleration
      - Dynamics of coronary blood flow
      - Study of human motion (Gait analysis)

   b. **Physiologic Modeling**: Use of engineering models and computer simulations to develop an understanding of physiological relationships.
      - Graphical computer modeling of head and neck response to acceleration
      - Analog models, 3D printing, cadavers, manikins
10. **Clinical Engineering**: Ensuring medical systems' integrity and regulatory compliance. Management of clinical facilities, devices, systems, and procedures.

In today's Intelligent Hospital, Biomedical Engineering will be required to support the overall broad base of technical disciplines outlined. The focus will be very institution dependent as well as based on an organization's existing organizational structure. Depending on the structure, strong interdisciplinary or interorganizational synergies will need to exist. Further in this section, an example of an organizational structure for the Biomedical Engineering department will be outlined which enables the technological design, development, and management operations to be effectively structured to meet the needs of the evolving Intelligent Hospital.

Historical Biomedical Engineering in Healthcare

The exact discoveries and their sequence that have defined Biomedical Engineering are arguable. However, if we examine the early discoveries and medical systems designs historically deployed, we can start to trace the origins of Biomedical Engineering within the healthcare environment and their impact on today's medical/biomedical systems in the Intelligent Hospital, as outlined below.

■ Artificial Pacemaker (Dr Albert Hyman)	1932	Figure 4.18
■ Heart Lung Machine (Dr Gibbons)	1937	Figure 4.19
○ Use in Human Surgery	1953	
○ Approved as Surgically Safe	1960	
■ Artificial Heart (Dr Willem Kolff)	Patented 1957	Figure 4.20
○ First Implants Drs DeBakey and Bernard	1966, 1967	
○ Jarik 7 Heart Dr Devries	1982	Figure 4.21
▪ Patient survived 112 days		
■ Artificial Kidney Dr Abel	Patented 1913	Figure 4.22
○ Dialysis Dr Kolff		

Figure 4.18 Hyman pacemaker.

Figure 4.19 Dr Gibbon's Heart-Lung machine.

Figure 4.20 Kolff artificial heart.

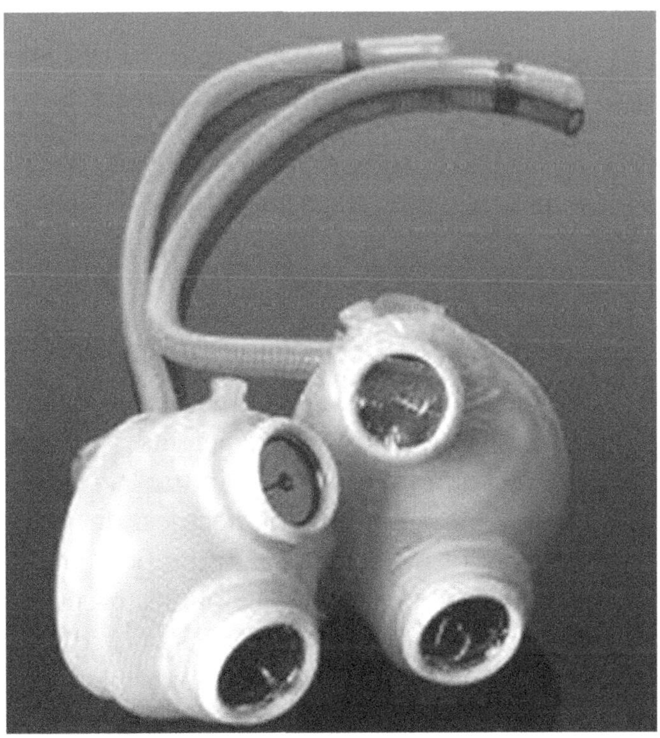

Figure 4.21 Jarvik artificial heart.

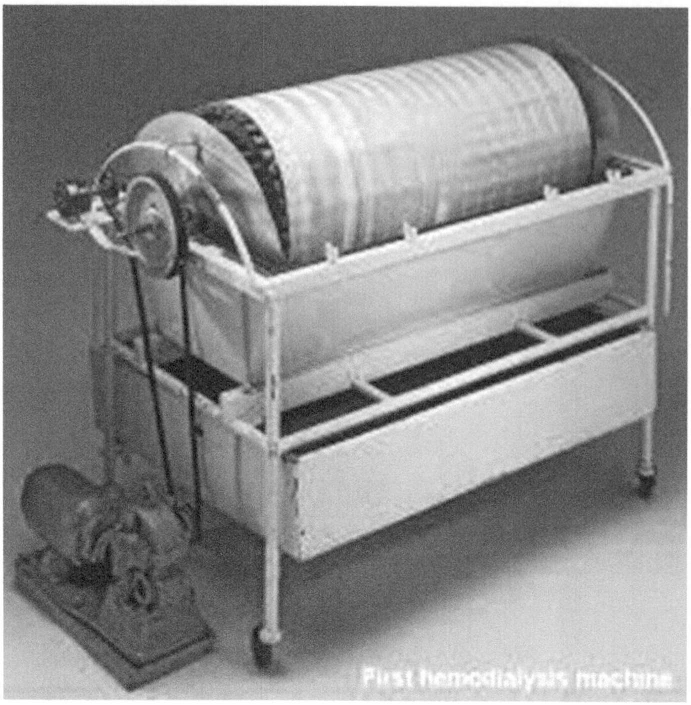

Figure 4.22 Dr Abel first hemodialysis system.

Creation and Evolution of Clinical Engineering

In the early years of hospital-based Biomedical Engineering operations (circa late 1960s through 1975), the roles and responsibilities varied and were typically disjoint and performed by individual hospital departments to ensure specific solutions for instrumentation, workflow, and the maintenance and repair of medical devices. There did not exist any standard approach to equipment management or rigid regulatory requirements. In most cases, medical devices and hospital systems were directly supported by the device manufactures under a variety of support agreements.

As summarized by Malcolm Ridgeway [27] in the late 1960s, concerns started to develop around the grounding of electrical devices used in the direct care or management of patients through the potential to introduce micro-electric shocks that could result in patient electrocutions. In 1971, Ralph Nader, a lawyer and consumer activist, was quoted in an article published in the Ladies Home Journal, stating that a minimum of 1200 Americans were electrocuted annually during routine diagnostic and therapeutic procedures [28].

This publication and report started the sequence of events which drove the beginning of a new discipline of engineering within hospitals and other healthcare areas to support medical device management and ensure patient safety concerns commonly referred to as Clinical Engineering, Biomedical Engineering, or Health Technology Management. Clinical Engineering departments focused to centralize and formalize a medical device testing program within all the hospitals in the United States. Functionally, this started as electrical safety testing of devices and eventually expanded to provide other functions including preventative maintenance, calibration, repairs, and overall device upgrades and management. Today, most hospitals still test for electrical safety as part of their device acceptance testing procedure and commonly following major device repairs.

The Traditional Mission of Clinical Engineering (Health Technology Management)

Over the many years since the start of Clinical Engineering, the role and operations have significantly expanded beyond simple electrical safety testing and verifications. The fundamental concept and purpose of Clinical Engineering has focused on the execution of an institutional "Medical Equipment Management Program" targeting operational integrity and compliance with the Joint Commission (TJC) standards and guidelines to ensure regulatory compliance and accreditation of the healthcare organization. The Joint Commission is a private, non-profit organization dedicated to improving the quality of care within organized healthcare organizations which includes in-patient hospital operations, outpatient care in the organization's regional sites, physicians' offices, and home health and telemedicine operations. The TJC provides a comparative measurement between hospitals and healthcare organizations, fulfilling State licensing requirements, and provides metrics to fulfill Medicare and HMO requirements. Accreditation performance can directly influence an institution's charge rate and impact on liability insurance. The Joint Commission establishes confidence in hospital operations and ensures that all hospitals and organizations operate at required standard levels. TJC examines overall hospital operations and in terms of Clinical engineering focuses on the processes associated with medical equipment management ensuring operational integrity and patient/staff safety.

The Medical Equipment Management Program can be defined as follows:

> **Medical Equipment Management Program (MEMP)** includes the business processes used in interaction and oversight of the medical equipment involved in the triage, diagnosis, treatment, and monitoring of patients. The related policies and procedures govern activities such as the selection, planning, and acquisition of medical devices. Incoming inspection and acceptance testing, PM, maintenance, and eventual retirement and disposal of medical equipment. Medical equipment management is a recognized profession within the medical logistics domain. The healthcare technology management professional's purpose is to ensure that equipment and systems used in patient care are operational, safe, and properly configured to meet the mission of the healthcare; that the equipment is used in an effective way consistent with the highest standards of care by educating the healthcare provider, equipment user, and patient; that the equipment is designed to limit the potential for loss, harm, or damage to the patient, provider visitor, and facilities through various means of analysis prior to and during acquisition, monitoring and foreseeing problems during the lifecycle of the equipment, and collaborating with the parties who manufacturer, design, regulate, or recommend safe medical devices and systems.

Compliance with the guidelines, TJC looks for consistent operation defined through the policies and procedures as outlined in Appendix 1, which is supported through a consistent set of documentation. The specific functions and objectives of the program can be outlined as follows:

- Medical Equipment Selection and Specification
- Facilities and Installation Requirements
- Inventory Management and Control
- Unique Identification
- Incoming Acceptance testing
- Risk Assessment and Classification
- Critical or Non-Critical Based on Risk to the Patient in the Event of Failure
- Preventative Maintenance Management
- Development of PM Procedures

- PM Interval Determination
- Service and Maintenance
- General Repairs and Service Methodology
- Calibration
- Battery Replacement
- Device Recalls and Hazard Management
- Incident Management
- Documentation and Service History
- Ownership Information
- Service Documentation and History
- Reporting to Executive Groups Addressing Safety and Risk Management
- Annual Evaluation and Training Standards

The equipment management program is designed to ensure that medical device functionality, operational integrity, patient safety, and regulatory compliance are continuously reviewed, evaluated, maintained, monitored, measured, and achieved. Clinical Engineering continues to remain a vital and critical component of the hospital accreditation process and is a vital component of the broader scope and role of Biomedical Engineering operations in today's Intelligent Hospital.

Over the last decade, with the increasing advent of the Intelligent Hospital concept representing increased device integration, connectivity, establishing a seamless data environment, the Equipment Management Program has transitioned to a broader definition commonly defined as the Technology Management Program including expanded definition of medical devices, including clinical software, IoT devices and supporting technologies such as RFID, RTLS, BLE, AI, etc.

Medical Systems, Devices, and Equipment

The FDA definitions of a medical device encompass a broad set of categories as outlined.

Medical Device [29, 30]

A medical device is defined as any instrument, apparatus, implement, machine, appliance, implant, reagent for in vitro use, software, or complex systems, used alone or in combination used for the prevention, diagnosis, treatment, monitoring, and management of a disease or physical injury in humans and animals.

Medical Equipment represents a subset of medical devices defining the support, management, and inclusion into a hospital MEMP. These are typically electronic devices capable of transmitting, receiving, and storing a dataset including PHI, clinical parameters, and dynamic metrics as defined below.

Medical Equipment

Medical Equipment is defined as medical devices requiring repair, a combination of preventative maintenance, calibration or performance verification, user training, and decommissioning. **Medical Equipment** includes all systems and devices used in the prevention, treatment, diagnosis, monitoring, direct care, and management of disease and injury in humans. This includes but is not limited to life support devices (such as ventilators), monitoring devices (such as physiological monitors), treatment devices (such as linear accelerators), diagnostic systems (such as imaging systems (CT, MR, ultrasound)), laboratory systems (such as hematology and chemical analyzers), communications systems (such as Nurse Call System), and Software/Systems that electronically transfer, store, convert according to preset specifications, or display medical data.

Within the hospital, patient medical systems or equipment are further categorized into several distinct classifications which are included in the MEMP program and specifically identify items that are excluded from the program. These are defined by the institution and reviewed by the hospital's safety, quality, regulatory, and legal organizations as follows:

Patient Care Devices: All devices within the hospital or patient environment of care directly used for triage, diagnostics, treatment, decision support, and general patient management.

Temporary Devices: These devices are used to support patient care providing similar functionality that are commonly non-hospital owned medical devices classified as loaners, rentals, patient owned, or third party owned devices. These devices are equally subject to regulatory compliance issues based on the individual policies and procedures of the hospital. (Refer to Appendix 2.)

Program Exclusions: Based on individual hospital policies, these devices are omitted from the medical equipment management plan. These can include implantable Devices, Disposable (Single Use) Devices, non-electric devices such as surgical instruments, commercial entertainment devices, and cell phones, iPads, etc.

These devices are managed in a variety of ways in accordance with institution's policies under the guidelines of the CMS, Joint Commission, and/or other regulatory agencies. The management and responsibility of these devices continues to be an arguable point of discussion on many levels between regulatory agencies, CMS, professional organizations, and hospital quality and safety groups. More recent concerns of cybersecurity attacks and potential security and PHI vulnerability associated with these devices have added a new dimension to these discussions which are addressed in other sections.

Medical System Equipment (Asset) Inventory Management and Control

Fundamental to the management of a broad base of medical devices located throughout the hospital is establishing an accurate inventory management program. Medical device information is most effectively maintained through a robust inventory database (CMMS) that uniquely identifies the devices and device characteristics and supports documentation, coupled with periodic validation processes.

Inventory Documentation (Data Fields)

The database (CMMS) is commonly the reference or library of all medical device information, defining all device characteristics, including unique identification, maintenance requirements, previous services, calibrations, preventive maintenance, upgrades, ownership, locations, etc.

To clearly identify and report on the current and historical status of the medical devices, the database defines all device characteristics through a defined set of data fields. The selection and data content vary by institution and are dependent on the needs of the hospital; however, regulatory agencies do expect a specific and consistent set of information to be readily available for each medical device. A comprehensive representation of a typical dataset maintained by today's Intelligent Hospital as a function of device is defined in the following sections.

Initially, clinical engineering asset management systems have focused on regulatory compliance and devices specifically used in the care and management of patients and defined as part of the Medical Equipment Management Program. In today's Intelligent Hospital, the content represents a much broader set of device types, including enabling devices such as handheld devices, dashboards, RTLS, BLE devices, clinical software applications, network hardware, communication devices, and servers. The types of devices and associated information contained within the database are dependent on the scope and specific responsibilities, deliverables, objectives, and goals of the institution and individual departments.

Overview: Medical Device Data Content

The data content associated with medical devices is driven by many factors and by definition is the archive of all information associated with a medical device. This can include manufacture information, ownership, business continuity data, operational data, cybersecurity, financial information, etc. The content is defined by the Intelligent Hospital management and can be all-encompassing. Highlighted below and detailed in Appendix 2 are the parameters that represent a potential comprehensive dataset that would be required by the Intelligent Hospital which is highlighted below.

- Medical System/Device Description and Operational Details
- Regulatory Compliance Information
- Institutional or Department Ownership Details
- Financial Data
  - Capital Replacement
  - Purchase Cost
  - Dates of Purchase, Life Expectancy, End of Date Estimate
  - Service Cost
- Inventory Type
  - Medical device, enabling technology, communications devices, etc.
- Unique Device Identifier
  - ID Number, barcode, QR code, Passive or Active RFID, BLE tag

Note: In the Intelligent Hospital, a combination or all identifiers would typically be used simultaneously to support many operational scenarios and processes.

- Device Classification
  - Medical, Temporary, Software Application, Enabling Technology, Other
- Device Type
  - Defibrillator, Anesthesia, Linear Accelerators, Infusion Pump, etc.
- Device Name

- Device Specific Information
  - Manufacture, Make, Model, Serial Number
- Device Status
  - Active, Retired/Decommissioned, Out for Service, or Repair
- Usage:
  - Patient Care, Research, Home Care
- Medical Device Class:
  - FDA Definition Class 1, 2, or 3
- Risk Assessment and Level: Patient Risk Assessment
- Network Specific Data
  - Networked/Wireless
- Software System/Firmware
- Service Policy
  - OEM or AEM
- Ownership
  - Executive and Technology Owners, Department, Dedicated User
- Service Methodology
  - Internal Department
  - OEM
  - Service Contract
  - Warrantee
  - Insurance
  - Time and Materials
- Regulatory Requirements
  - Calibration, Preventative Maintenance
  - Laser Safety Requirements
    - Lasers Class, Wavelength, Output Power
- Location and Tracking
- End of Life Planning
  - Purchase Date, Replacement Cost, End of Life
- Cybersecurity Management
  - System Definition, Systems Level, Device Level
  - Vulnerability Risk Level
  - Vulnerabilities, Control Requirements

Health Technology Management: Service and Support Modalities

To effectively manage and support the board base of devices included in the Medical Equipment Program, clinical engineering responsibilities are typically distributed across multiple operating groups, requiring varying levels of engineering expertise and specific knowledge about the equipment to effectively support and manage the devices as outlined:

• High Energy Devices • Radiological/Imaging Devices • Laboratory Systems • Patient Care and Management Devices • Regional Network Support	(Biomedical Electronics Group) (Radiological Engineering) (Clinical Engineering) (Clinical Engineering) (All of the Above)

Health Technology Support provides equipment management functions for Patient Care devices across all clinical departments throughout the hospital, off-sites, and regional network facilities ensuring that patient care equipment follows regulatory guidelines and is standardized across all medical facilities and locations. This includes device repairs/service, preventative maintenance, recall management, device incident management, and remediation processes. These services are typically provided through a combination of service modalities, internal staff service, time and materials via third party or OEM, and service contracts. The distribution between these service modalities can vary based on the size of the institution, the number of devices, and the administrative philosophy. In a large academic institution, typically 50–60% of the devices are serviced by internal staff and 40–50% are serviced on a contracted service agreement, either time and materials (T&M) or service contract. This overarching group provides direct maintenance and oversight of devices/systems service by OEM (original equipment manufacturer) or alternate service methodologies, including internal service or third party).

Clinical Engineering Group provides the traditional hospital role of maintaining and supporting the in-patient and outpatient clinical equipment, such as defibrillators, pumps, blood warmers, physiological monitoring, etc. The primary responsibility of this group focuses on the management of the overall clinical equipment service program providing the direct in-house maintenance and support of day-to-day equipment services, including inspection, preventive maintenance, repair, calibration, safety assurance, pre-purchase comparative equipment analysis, and incident investigation. In addition, the group manages the equipment and provides oversight. As in most organizations, the group participates in a variety of Hospital committees and ensures regulatory compliance regarding the management of medical equipment and devices.

Biomedical Electronics provides equipment management and direct technical support to Radiation Oncology, providing management, repair, and preventive maintenance to ensure regulatory compliance of high energy devices including linear accelerators, proton therapy systems, brachytherapy systems, simulators, and supporting equipment. These systems can be classified as high energy delivery systems requiring significant engineering and physics expertise and specific device training. In many institutions, these devices can be directly under service contract; however, in some dynamic institutions, internal staff support device service and preventive maintenance via shared risk agreements.

Laboratory Systems provides equipment management and direct technical support to Laboratory medicine, providing repair and preventive maintenance of laboratory devices and supporting equipment. Similarly, staff associated with the service and PM of these devices require enhanced expertise and training.

Regional Network Support provides equipment management and support located at remote locations or regional ambulatory care areas of the Intelligent Hospital. However, business and management processes to support remote locations require different workflows to effectively provide these services. The standards and guidelines remain consistent with the overall management policies. Since these regional locations can be very focused on specific patient care or treatment delivery processes, management of service becomes key to effectively execute support. Dedicated staff from other groups are required to address very specific device issues.

Service Documentation

Regulatory Compliance, as well as good business practices, requires all hospitals to maintain a comprehensive record of all services and critical events associated with each of the devices in the program as outlined below:

- Documentation of Service (Procedures or Checklists)
- Incoming Inspections/Acceptance Testing Prior to Use
- Work Orders of Services/Repairs
- Documentation of Upgrades (Hardware and Software)
- Calibrations and Preventative Maintenance
- Recall Services
- Patient/Staff Incidents
- Decommissioning
- Cybersecurity

Dynamic Dashboard and Reporting

Ideally, the Intelligent Hospital provides real-time status reporting on the state of medical equipment within the Institution. There are a variety of reporting mechanisms possible within the database representing a set of dynamic, real-time configurable reports and visual dashboards capable of providing dynamic alerts and alarms for managing operational integrity and business continuity.

Most hospitals have introduced varying levels of hospital intelligence and started an ongoing transition toward a formalized Biomedical Engineering operation having synergy and designed to support advanced technologies, integrations, and advanced communications. These capabilities will be required to develop, manage, and support enhanced hospital intelligence. Multiple development, management, and support systems will be required to provide the traditional regulatory compliance objectives coupled with new objectives and deliverables of the Intelligent Hospital.

Advanced Biomedical Engineering Functionality in the Intelligent Hospital

Over the last several decades, the changing healthcare environment has driven hospitals to critically evaluate and optimize their operations to enhance patient treatment and care while focusing on increasing financial constraints. The hospitals have moved to support an increasing outpatient care environment, driving an increase in in-patient acuity level. Hospitals have focused on new and evolving technologies to enhance patient care, ensure safety, and optimize workflow and operational efficiency, which are the defining characteristics of the Intelligent Hospital.

A critical problem that exists in most hospitals, which directly impacts and limits the development of an integrated Intelligent hospital, is that the deployment of advanced technologies and medical systems is not well organized or planned and typically occurs over long periods of time. In addition, these advancements are frequently sponsored or focused through very specific applications or departments and not necessarily focused on institutional solutions. This approach limits considerations for broader based solutions and standardization and creates silos of technology targeting unique specific functional solutions. The transition to a seamless patient care model and associated care environment requires a strategic institutional plan and focuses on creating institutional solutions, where possible, coupled with a broad set of enabling technologies as discussed in Section 1. As highlighted, any development and deployment plan must include a parallel supporting and management plan and methodology providing a broad set of expertise and synergies that cross the typical department boundaries and structure existing in most hospitals.

The Intelligent Hospital concept focuses to establish a seamless integration of many diverse technologies, including medical devices, wireless sensors, clinical applications, as well as the integration of location context provided by RFID and RTLS technologies, IoT device, and new applications including analytics, medical record, prediction, and AI all integrated within clinical scenarios. This vision enables seamless sharing of information across all the patient care and management modalities, starting at patient triage, diagnosis, all the way through specialized treatment and recovery.

The evolution to the "Intelligent Hospital" significantly alters and expands the technical role, functionality, and operational responsibilities of Biomedical Engineering to become increasingly adaptive with an increasingly focused expanding role which includes basic research, investigating new and evolving technologies, design and development, analytics and informatics while continuing the traditional responsibilities associated with health technology and medical device/systems management ensuring regulatory compliance.

Since medical devices are no longer stand-alone devices, now networked, integrated with clinical applications and analytics, and commonly support wireless capabilities these complex systems are increasingly susceptible to infrastructure and cybersecurity issues that potentially impact operations and patient safety.

To effectively manage and support these complex systems advanced skills as well as developing facilities, such as a simulation laboratory to dynamically test, evaluate, update, and service medical systems, test new applications, assess and test cybersecurity vulnerability, and device life cycle impacts will be required. Biomedical Engineering must expand its traditional expertise and skill sets to manage these new responsibilities to support the deployment of evolving technologies and integrations with medical devices with enabling technologies such as RFID/RTLS, Low Energy Bluetooth, advanced software applications, and other wireless commercial devices. The following highlights the new functionality and synergies necessary for an Advanced Biomedical Engineering and technology management operation within the Intelligent Hospital.

Expanded Biomedical Physics and Engineering Role and Functionality

- **Traditional Health Technology and Medical Equipment Management**
  - Ensures medical operational integrity and regulatory compliance
  - Maximizes patient safety and satisfaction
- **Electronics and Instrumentation**
  The design and development of sensors, surgical tools, and electronics to monitor and measure physiologic events and enable advanced treatments
- **Mechanical Engineering**
  Develop patient specific and centric custom solutions
  - CAD/3D Printing
    - Anatomically Accurate Models and Visualization
  - Patient Specific Solutions
    - Shielding and Immobilization Device
    - Targeting Systems for Radiation Therapy
  - Development of Custom Surgical Devices, Cutting Guides

Note: The operational components of Mechanical Engineering, including the mechanical development shop, 3D printing laboratory, and the shielding and immobilization fabrication facility, as described in Section 3.

- **Biomaterials**
  The study, design, and development of biocompatible implantable materials to support reconstructive surgery and anatomic replacement materials
- **Imaging**
  The development of graphical visual interfaces visualizing detailing human anatomy as well as physiologic functions.
  - 3D Printing
  - AR/XR Visualization
- **Computer Science and Software Development**
  Developing analytic applications providing integrations and visualization to support data management and applications
- **Cybersecurity**
  Enable Vulnerability Assessment, development of security controls to minimize vulnerability, and the development of policies and procedures to ensure operational integrity and system recovery

- **Medical Informatics**
  Management of patient-related data interpreted results supported through decision tools to assist in the clinical decision support process. These systems range from simple database systems to expert systems AI driven and neural networks
- **Research and Investigative Studies**
  - **Biological Effects:** The study of the effects and consequences of electrical and magnetic fields on biological tissues and their functions
  - Impact of Electromagnetic Interference
  - **Biomechanics:** Study of dynamic mechanics associated with Physiologic Systems
  - Study of head and neck response to transitory acceleration
    - Dynamics of coronary blood flow
    - Study of human motion

In conclusion the advancement of the Intelligent Hospital necessitates that Biomedical Engineering transform well beyond the traditional support roles and evolve into a dynamic contributor focused on clinical patient care and treatment operations driving the design and development of unique and patient specific clinical solutions, processes, and workflows which directly impact and enhance patient outcomes, safety, satisfaction, and overall care. To enable this evolution of Biomedical Engineering, hospitals must redefine their organizational structure, focusing on and enabling high degrees of integration and collaboration.

The Advanced Biomedical Physics and Engineering Organization

At MSKCC, Biomedical Engineering has been part of the Medical Physics Department providing direct support to Radiation Oncology patients undergoing radiation therapy since the 1990s. Medical Physics is a clinical department focused on the development and enhancement of techniques for utilizing ionizing radiation, ultrasound, lasers, thermal, magnetics technologies, imaging technologies and computation analytics for the diagnosis and treatment of cancer. This represents an ideal fit for the inclusion of an advanced and adaptive Biomedical Engineering program providing opportunities to exploit a broad set of engineering capabilities to further target and enhance the effectiveness of targeted cancer treatment. Cancer treatment utilizes a combined approach to manage cancer care, commonly involving a combination of surgery, chemotherapy, and radiation therapy.

Surgery focuses on removal of a localized cancer mass or tumors and chemotherapy provides a broader systemic approach to attack cancer targeting multiple locations simultaneously. Radiation therapy targets localized tumors using ionizing radiation to damage cancer cells which leads to the eradication of the tumor. Radiation therapy is delivered in one of three ways: direct exposure to beams of photons and charged particles from accelerators; brachytherapy where small radioactive sources are implanted or inserted into the body where short-range gamma-ray and beta-particle radiations which kill cancerous cells while minimizing damage to normal tissues; and systemic radiation delivery where a radioactive labeled pharmaceutical is administered that seeks out the target cells. The objectives of a radiation oncologists and medical physicists are to develop techniques to maximize the effects of radiation treatment while minimizing the damage to the surrounding tissues. To support these objectives, Medical Physics provides the integrated and coordinated functions of clinical physics, radiation treatment planning, brachytherapy, spectroscopic and MR physics, nuclear medicine, radiation and laser safety, and Biomedical Engineering.

Today, Biomedical engineering has incorporated capabilities to fabricate patient specific radiation shields (blocks) which mount directly to the Linear accelerator to block or minimize radiation exposure to out of field anatomy of the patient. Biomedical Engineering contributions have significantly expanded focusing on direct patient treatment modalities, including 3D imaging reconstruction, development of 3D anatomic accurate models to enhance radiation targeting applications, as well as dynamic treatment planning using AR/XR techniques. Enhanced Radiation therapy requires the minimization of radiation exposure to surrounding tissue, which focuses on customized patient specific shielding devices as well as immobilization devices to minimize movement and enhance targeted positioning and delivery.

An effective approach adopted at Memorial Sloan Kettering was to establish an enhanced Biomedical Engineering department of section operating incorporated as part of the Medical Physics Department organized and structured into multiple operating units focused on specific functionality and requirements outlined below and illustrated in Figure 4.23.

1. **Enhanced Health Technology Management (Clinical Engineering)**
   a. Provide the traditional service role for medical systems and regulatory compliance.

2. **Design and Development**
   a. Investigation of new and evolving technologies focused on the integration to provide enhanced clinical operations, applications, and workflows.
   b. Design Patient Specific/Centric Medical Solutions
   c. System Integration
   d. Software Development
3. **Advanced Project and Business Management Supporting**
   a. Cybersecurity Medical Devices and Systems
   b. 3D Imaging and Printing Laboratories
   c. Clinical Simulation: Supporting clinical operations and education, testing new software and technologies, and cybersecurity vulnerability and recovery procedure development.
   d. Business and Financial Planning
4. **Research**
   a. Participate in basic research, research programs, test and evaluation and educational programs.

In conclusion, to outline the operations and direct impact on treatment delivery and patient care and safety, the following case study provides an overview of multiple contributions from Biomedical Engineering to a single patient treatment process.

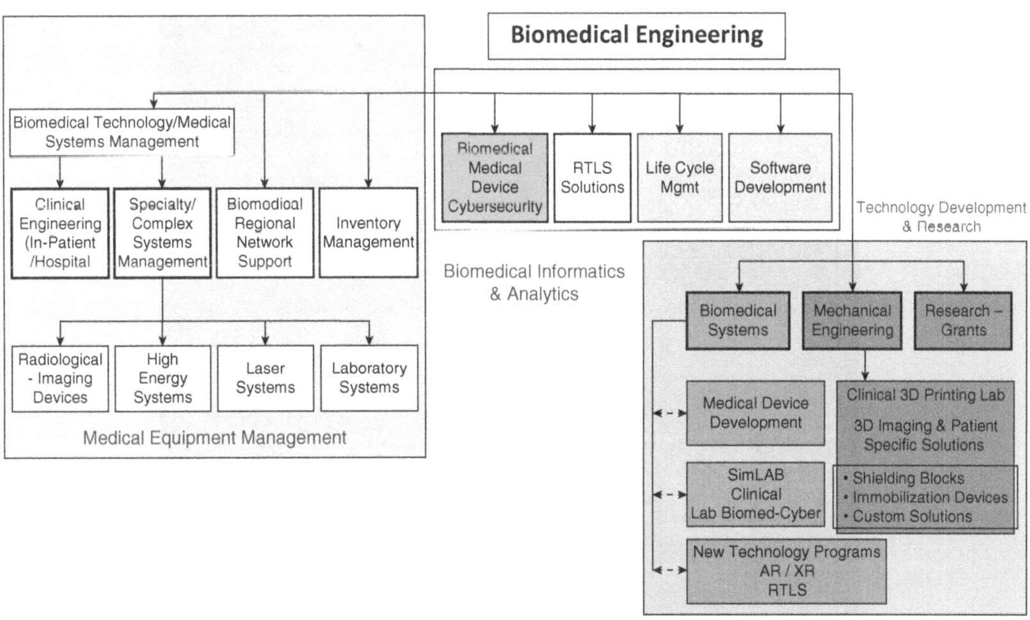

Figure 4.23 Advanced biomedical engineering organizational structure.

Case Study: Patient Treatment Scenario (Optimizing Radiation Therapy Effectiveness)

As an example, Biomedical Engineering's clinical role and contribution in patient radiation treatment scenarios includes a variety of patient specific devices or solutions to enhance radiation treatments and patient safety. In summary, Figure 4.24 illustrates a custom anatomically accurate 3D printed hip identifying a tumor location. Examination of the imaging and other forms of visualization (anatomically accurate 3D print, AR/XR visualization) a unique treatment plan is developed to ensure the optimum radiation dose as a function of position is delivered to the site of the tumor. Radiation delivery to the hip tumor may require patient specific shielding, such as the testicular patient specific shields (designed to protect the sensitive areas from unnecessary radiation exposure) (Figure 4.25), coupled with a patient specific immobilization device to ensure consistent positioning maximizing targeted radiation exposure of the patient (Figure 4.26).

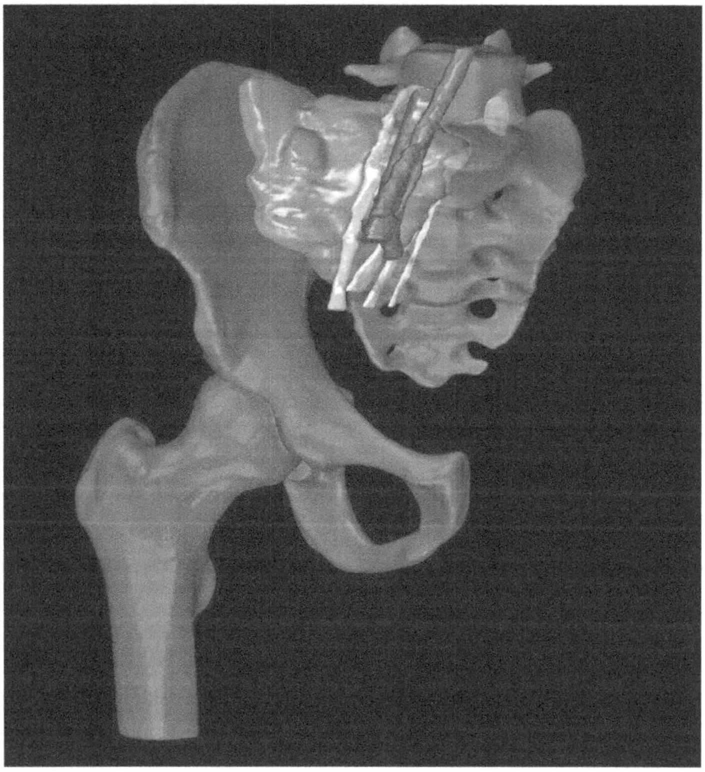

Figure 4.24 Advanced visualization: anatomically accurate 3D printed model hip.

Figure 4.25 Testicular radiation shield.

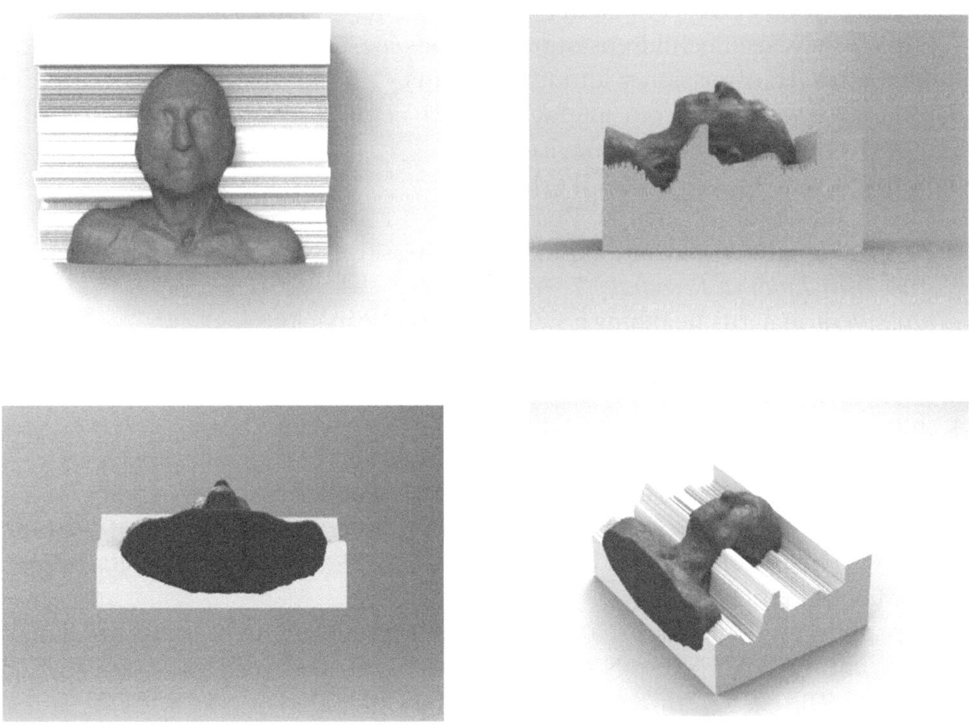

Figure 4.26 Anatomically match immobilization device.

As discussed in Section 3, many custom mechanical devices are designed to create patient specific shields, immobilization for an optimum radiation dose delivery.

References

1. Frisch PH, Pappas HP (2023). Leveraging Technology as a Response to the COVID Pandemic, Routledge, ISBN 97803677693, Chapter #2: Optimizing Infection Control & Hand Hygiene during COVID.
2. The Editorial Board (2020). "Is the world ready for the coronavirus?—Distrust in science and institutions could be a major problem if the outbreak worsens". *The New York Times.* Retrieved 30 January 2020.
3. Page J, Hinshaw D, McKay B (26 February 2021). "In hunt for COVID-19 origin, patient zero points to second Wuhan market – The man with the first confirmed infection of the new coronavirus told the WHO team that his parents had shopped there". *The Wall Street Journal.* Retrieved 27 February 2021.
4. https://www.worldometers.info/coronavirus/
5. McNeil Jr DG (2 February 2020). "Wuhan coronavirus looks increasingly like a pandemic, experts say". *The New York Times.* Retrieved 4 April 2020.
6. Medical News Today Newsletter, ZawnVilinette, 5 May 2020.
7. https://www.seattlechildrens.org/conditions/a-z/COVID-19-exposure-but-no-symptoms/Transmission without symptoms
8. https://www.nytimes.com/interactive/2021/us/new-york-COVID-cases.html
9. http:/worldometer.info/coronavirus
10. https://COVID.cdc.gov/COVID-datatracker
11. Chiam T, Subedi K, Chen D, Best E, Bianco FB, Dobler G, Papas M (2021). Hospital length of stay among COVID-19-positive patients. *Journal of Clinical and Translational Research.* **7** (3): 377–385.
12. https://coronavirus.health.ny.gov/hospital-bed-capacity
13. https://www.beckershospitalreview.com/patient-flow/2-healthcare-leaders-talk-icu-capacity-5-sound-bites.html
14. https://fox11online.com/newsletter-daily/what-you-need-to-know-about-icu-capacity-in-the-united-statesFoxNews
15. Wunsch H, Wagner J, Herlim M, Chong DH, Kramer AA, Halpern SD (December 2013). *"ICU occupancy and mechanical ventilator use in the United States". Journal of Critical Care Medicine.* 41 (12): 2712–2719. doi: 10.1097/CCM.0b013e318298a139
16. https://www.google.com/search?q=NYC+Hopsital+lenght+of+stay&rlz=1C1GCEA_enUS976US976&oq=NYC+Hopsital+lenght+of+stay&aqs=chrome.69i57j33i10i160j33i10i299.5912j1j7&sourceid=chrome&ie=UTF-8
17. Hunter A, Johnson L, Coustasse A (2014) *"Reduction of intensive care unit length of stay: The case of early mobilization". Health Care Manag (Frederick).* **33** (2): 128–135. doi: 10.1097/HCM.0000000000000006.

18. Vekaria B, Overton C, Wiśniowski A, Ahmad S, Aparicio-Castro A, Curran-Sebastian J, Eddleston J, Hanley NA, House T, Kim J, Olsen W, Pampaka M, Pellis Lorenzo, Ruiz DP, Schofield J, Shryane N, Elliot MJ (2021). "Hospital length of stay for COVID-19 patients: Data-driven methods for forward planning". *BMC Infectious Diseases.* **21**, Article number 700.
19. https://www.barclaydamon.com/alerts/COVID-19-nys-hospital-capacity-to-increase-amidst-unprecedented-regulatory-pragmatism
20. Grimms SL (2015 Spring). "Evolution of a new risk-based approach to effective heathcare technology management". *Biomedical Instrumentation Technology.* Suppl 34–42.
21. Ridgeway Malcolm: https://accenet.org/awards/Pages/MalcolmRidgway.aspx
22. Ralph Nader: https://en.wikipedia.org/wiki/Ralph_Nader
23. "Overview of Medical Devices and Their Regulatory." http://www.fda.gov/AboutFDA/CentersOffices/OfficeofMedicalProductsandTobacco/CDRH/CDRHTransparency/ucm203018.htm
24. "Maintenance Management for Medical Equipment", American Society for Healthcare Engineering of the AHA, 1996, section II 6-7.
25. https://COVIDtracking.com/data
26. Rosenbaum L (2020). "Facing COVID-19 in Italy—Ethics, logistics, and therapeutics on the epidemic's front line". *New England Journal of Medicine.* **382**: 1873–1875.
27. Cuomo A (2020). New York State governor's daily coronavirus press briefing. New York City.
28. Chen GH, Hellman S, Irie T, Downey RJ, Fischer G (2020). "Regulating inspiratory pressure to individualise tidal volumes in a simulated two-patient, one-ventilator system". *British Journal of Anaesthesia.* **125** (4): 366–368.
29. Everington K (2020). "Taiwanese doctor invents device to protect US doctors against coronavirus". *Taiwan News*, 23 March.
30. Formlabs, Inc., "3D Printed Test Swabs for COVID-19 Testing" [Online]. https://formlabs.com/COVID-19-response/COVID-test-swabs/

Bibliography

Frisch PH, Miodownik S, Booth P, Carragee P, Dowling M, Patient Centric Identification and Association. IEEE Engineering in Medicine and Biology Society, Proceedings of the 31th International Conference. 2009: 1722–1725.

Grant MC, Geoghegan L, Arbyn M, Mohammed Z, McGuinness L, Clarke EL, Wade RG (23 June 2020). "The prevalence of symptoms in 24,410 adults infected by the novel coronavirus (SARS-CoV-2; COVID-19): A systematic review and meta-analysis of 148 studies from 9 countries". *PLOS One.* ***15** (6):* e0234765. doi: 10.1371/journal.pone.0234765.

https://www.cdc.gov/coronavirus/2019-ncov/symptoms-testing/symptoms.html

"Interim Clinical Guidance for Management of Patients with Confirmed Coronavirus Disease (COVID-19)". U.S. Centers for Disease Control and Prevention (CDC). 6 April 2020. Archived from the original on 2 March 2020. Retrieved 19 April 2020.

Jump up to: a b CDC (11 February 2020). “Post-COVID Conditions”. U.S. Centers for Disease Control and Prevention (CDC). Retrieved 12 July 2021.
Oran DP, Topol EJ (January 2021). “The proportion of SARS-CoV-2 infections that are asymptomatic: A systematic review”. *Annals of Internal Medicine.* ***174*** *(5): M20-6976*. doi:10.7326/M20-6976.
“Symptoms of Coronavirus”. U.S. Centers for Disease Control and Prevention (CDC). 22 February 2021. Archived from the original on 4 March 2021. Retrieved 4 March 2021.

Appendix 1: Biomedical Engineering Technology and Medical Device Management Program Policies and Procedures

Overview

The objective of the Technology Management Program is to ensure and optimize patient care, safety, and regulatory compliance in the use of medical equipment and supporting technologies. The Medical Equipment Management Program (MEMP) subject to regulatory compliance is a subset of the overall program focused on all medical devices, systems, and software which have filed and been defined as medical device with the FDA and used to diagnosis, treat, monitor, and provide overall care of the patient.

Biomedical Engineering is responsible for the evaluation, specification, and planning of new biomedical systems and technologies, as well as the management of preventive maintenance, repair, device recalls, and medical device related incidents of clinical systems, devices, and medical equipment within the Medical Center.

Guidelines

The Technology/Medical Equipment Management Program adheres to the guidelines and recommendations of the Centers for Medicare and Medicaid

Services (CMS) [1], the Joint Commission (TJC) [2], the ECRI Institute [3], the FDA [4] and other regulatory agencies including the DOH, and other device specific organizations.

It is the requirement of the Intelligent Hospital (IH) that all equipment defined as medical and used within the clinical environment be:

- Included in the biomedical equipment inventory (Biomedical Equipment Management Database (DB)).
- Include a documented service history maintained for each piece of equipment.
- Manage preventative maintenance and performance verifications.
- Manage all medical device recalls and manufacturer alerts.
- Support patient care processes and response for patient-related incidents.
- Provide access and reporting of the status of medical device inventory.

Responsibilities

Departmental/User Responsibilities

It is the responsibility of all departments, equipment owners, and users to ensure that their medical devices are represented in the biomedical device inventory and that the information is accurately identified and that all service histories are documented, archived, and current.

Biomedical Engineering Responsibilities

Biomedical Engineering is responsible for setting guidelines for identifying, evaluating, and inventorying medical equipment to ensure that patients, staff, and visitors are safe from electrical hazards and equipment malfunction. This is achieved by establishing criteria based on function, use, risk levels, maintenance, and historical data of all patient care equipment. Specifically:

- *Inventory*: All patient care equipment (i.e., purchased, leased, rental, consigned, and privately owned) is included in the Biomedical Engineering equipment inventory. It is the responsibility of biomedical engineering to provide unique identification labels (i.e., ID number and/or RFID) for this equipment.

- *Risk assessment*: Biomedical Engineering manages medical equipment risks. Risk criteria or manufacturers' recommendations are used to determine preventive maintenance frequencies and procedures. Biomedical Engineering identifies critical (high-risk) medical equipment in the inventory for which there is a risk of severe injury or death to a patient or staff member should the equipment fail. All critical (high-risk) equipment that is in the inventory is identified as such. Note: Critical (high-risk) medical equipment includes life-support equipment.
- *Acceptance testing and preventive maintenance*: Biomedical Engineering performs and/or confirms pre-deployment acceptance testing, incoming inspection, and performance verification. Also managed is Preventive Maintenance compliance with particular emphasis on PM requirements (both PM intervals and procedures). These activities and associated PM frequencies are in accordance with either manufacturers' recommendations or strategies of an Alternative Equipment Maintenance (AEM) Program.
- *Repair, maintenance, and recalls*: Biomedical Engineering identifies the activities for maintaining and testing all medical equipment included in the inventory and also monitors and records recalls and incidents.

Biomedical Engineering, through the Medical Equipment Management database, provides access and reporting to all users as to the state of the devices. This includes user inventory, all services, PM, recall, incidents, etc. Biomedical Engineering disseminates information at minimum biannually to all users via the medical equipment management committee and reports to the Hospital Executive Safety Committee about the status and safety of medical equipment and the overall equipment management program.

Equipment Specification, Planning, and Purchase

Critical to the development of a strong Technology/Medical Equipment Management Program, Biomedical Engineering participates in short- and long-term planning, equipment specification, and comparative analysis between competitive products. Coordination of these efforts ensures compliance with hospital strategic and technology objectives, clinical needs, infrastructure, standardization, deployment plans, and regulatory compliance. Toward that goal, Biomedical Engineering solicits input from individuals who operate and service equipment when it selects and purchases medical equipment. This is a multidisciplinary process that is coordinated through an Institutional Clinical Device Product Evaluation. Consideration of clinical needs, cost-benefit analysis, performance, and safety is all considered. The ECRI Institute and any other organizations deemed necessary are used as resources.

All new medical equipment identified for use in patient care areas and defined as part of the equipment management program will be required to meet applicable safety standards. At the recommendation of Biomedical Engineering, Corporate Procurement, purchase requisitions will be reviewed for compliance with safety and maintenance standards.

Medical Equipment Management Program

Included are all devices, systems, and software defined as medical devices by the FDA, as well as supporting devices associated with the environment of care, devices used in research or documented for other business applications. All devices are categorized as either medical or non-medical.

Definitions

Medical Equipment is defined as all systems and devices used in the treatment, diagnosis, and monitoring of patients. This includes life support devices (such as ventilators), monitoring devices (such as physiological monitors), treatment devices (such as linear accelerators), diagnostic systems (such as imaging systems (CT, MR, ultrasound), laboratory systems (such as hematology and chemical analyzers), communications systems (such as Nurse Call System), and Software/Systems that electronically transfer, store, convert according to preset specifications, or display medical data and were filed with the FDA as medical device.

Support Methodology, all devices defined within the Biomed Database are supported through a combination of service methodologies to include internal support, original equipment manufacturer (OEM) service support, and third-party service support. These services can follow different maintenance strategies (defined below), which are classified as: OEM (Original Equipment Management), AEM (Alternate Equipment Management), and General Support.

OEM Program [5]

Equipment/devices which are managed according to the OEM recommendations for PM interval and Test Specifications and/or performance verification. The OEM must program include:

- Radiological Devices and Systems
- Laser Systems*
- New Medical Devices/Systems in the Institution

AEM Program [6]

All eligible equipment which can be maintained using alternate support strategies as defined by qualified staff (i.e., Biomedical Engineering) and reviewed by the hospital's Executive Safety Committee.

General Support

This includes medical devices having no direct risk to the patient with or specific preventative maintenance or performance verification requirements, other than general inspection or non-medical devices requiring service based on good business practices (i.e., devices not used in patient care).

Inventory Requirements

All equipment, including medical, non-medical, and research devices, are uniquely identified, maintained, and supported under one of the service programs, either internally through Biomedical Engineering, Engineering, Facilities Management, equipment manufacturers, or third-party service providers (whether under a service contract or a time and materials basis) are required to be included (inventoried) in the Technology/Medical Equipment Management Program. The inventory data record contains all information required to support and manage the devices and is assigned a unique device identifier, manufacturer, risk level, and preventative maintenance requirements. Biomedical Engineering evaluates new types of equipment before initial use to determine whether the equipment should be included in the inventory. All devices receiving an incoming inspection

---

* High energy-based laser systems (3b and 4) have additional requirements imposed through the Institutional Laser Safety Program.

will be evaluated to determine if they are to be included in the Technology/Medical Equipment Management Program.

Service Documentation

Each device, system, or software application identified as a medical device is required to maintain a comprehensive service history documenting all inspections, repairs, preventative maintenance, software upgrades, modifications, recall-related services, reported incident responses, and general changes in device status.

Institutional Medical Equipment Database

The Technology/Medical Equipment Management Program is supported through a centralized database, containing the equipment inventory and all records of service, repair, preventative maintenance, and service cost. The database is available on the intranet and dynamically represents the status and history of all devices. The system automatically monitors and reports PM compliance and various performance statistics, including user errors, equipment uptime (where applicable), and service cost. The inventory contains equipment, devices, systems, and software broadly defined as medical and non-medical. These devices include clinical devices used in patient care, medical software, laboratory systems, both clinical and research, general equipment, and temporary devices.

It is the responsibility of equipment managers and users to periodically review, validate, and update the information in the database to ensure the accuracy of the equipment inventory, including timely entry of preventive maintenance and service histories in the database. Database training and review sessions are available through Biomedical Engineering upon request.

Medical Systems

Medical Equipment, as previously defined, includes all systems, devices, and medical software, defined as medical devices by the FDA used in the treatment, diagnosis, monitoring, and general patient care.

Non-Medical Equipment is defined as systems used in any area of the medical center not directly used in any of the above patient scenarios. These devices typically have no direct patient contact and include facilities-related devices and general equipment.

General Devices are devices which, without maintenance other than normal user care, are reasonably free from avoidable hazard. These devices can be medical or non-medical devices and are exceptions to the Technology/Medical Equipment Management Program. These devices are reviewed by Biomedical Engineering and committee defined by the institution's policies.

Non-Hospital Owned Equipment

Rental Equipment/Devices

Rental equipment can arrive at the hospital on a 24×7 basis, with some devices having a complexity beyond the scope of Biomedical engineering's expertise. Therefore, the vendors/OEM certification that the equipment is provided in proper working order and complies with all electrical safety requirements and PM compliance is required and acceptable.

Loaner/Demonstration Equipment

All loaner or demonstration equipment used in the patient care environment shall comply with the requirements of the Technology/Medical Equipment Management Program, including temporary identification and incoming electrical safety test. The vendor must provide certification that the equipment is working within performance specifications.

Patient Owned Devices

All patient owned medical equipment used by the patient in the hospital is required, at a minimum, to undergo an electrical safety inspection and will be provided a temporary inventory number. It is the responsibility of the patient or hospital staff to inform and notify Biomedical Engineering about the use of patient owned devices and request an inspection.

Non-Hospital Owned Third Party Patient Care Devices

In cases where the Hospital uses Non-Hospital Owned Third Party Patient Care Devices, such as drug delivery and infusion systems, it is the responsibility of the Third Party to manage and ensure regulatory compliance of all devices.

It is the responsibility of the Third Party to ensure all PM compliance is current and documented, ensuring device care and management, including disinfection, routine cleaning prior to redistribution. The third party's disinfection procedure should be made available to the MSKCC Infection Control Services. If the third party's disinfection procedure changes, the Infection Control Services should be notified. Equipment incidents, recalls, and hazard alerts are handled by the third party in accordance with the regulatory standards and documentation. All required device recalls or related upgrades and notifications are handled by the third party. It is the responsibility of the third party to ensure that all new or incoming equipment delivered to MSK has undergone a performance verification performed by a qualified service provider prior to use. The third party is required to maintain service records available to MSK at our discretion indicating incoming performance verification has been performed; audits will be periodically performed.

Non-Hospital Owned Personal Devices

High Power Devices

These are non-medical devices, which are high energy consumption devices, such as hair dryers, curling irons, clothing irons, toasters, hot pots, Foreman grills, electric razors, and hair trimmers. **These and similar devices are disallowed for use anywhere in the medical center.**

Consumer Electronics and Devices

These are typically entertainment devices that run directly on AC power, or through a DC adapter, such as laptop computers, iPods/iPhones, MP3/4 players, portable video players, and cell phones. These devices are allowed for general use within the hospital without safety testing. The use of these devices in specific patient care areas is at the discretion of the medical staff to ensure a quality patient care environment for all patients.

Cell Phones

The use of cell phones is prohibited in critical care areas (i.e., ICU) and telemetry areas. They are generally allowed to be used in other areas of the Hospital at the discretion of the clinical staff.

Conditions for Use (Non-Hospital Owned Devices)

- All devices used within patient care areas can be allowed or disallowed at the discretion of Biomedical Engineering or medical staff, to ensure the quality of the overall patient care environment for all patients.

- No device can use the red emergency power outlets, other than personal medical devices at the discretion of the medical staff.
- Extension cords cannot be used on any category of device or anywhere in the medical center other than in case of emergency [7].
- Power strips may be used when following TJC policy [8].
- Cell phones can be used in compliance with the Hospital's cell phone policy [9].

Devices other than those outlined within this policy will be reviewed and classified at the discretion of Biomedical Engineering. Any questions concerning the specifics of this policy or specific devices can be referred to Biomedical Engineering.

Unique Identification and Inventory

It is the policy of the Intelligent Hospital that all devices, consistent with the above definitions, be included in the Equipment Management database, including inventory and service history information. It is the responsibility of each user department to notify Biomedical Engineering of new or changing inventory. Biomedical Engineering performs a comprehensive inventory process and confirms that an incoming inspection or performance verification was performed prior to deployment into operational use.

It is the responsibility of the user department to notify Biomedical Engineering of new devices and ensure that this inventory and inspection or performance validation process is completed prior to use.

Equipment Labeling

Once a device has been inventoried into the database, a unique bar-coded inventory number and label are issued by Biomedical Engineering, as illustrated below, and attached to the device. This unique number becomes the primary identifier for the equipment within the database and must be included on all service documentation and equipment references. To enhance the management of the devices within the program and reduce missing or lost devices, additional identifiers such as RFID and BLE can be used and associated to the unique identifier.

Unique Inventory ID number: Including Barcode and QR Code Identifier

100000000

0 123456 7890

PM Confirmation Labeling

Biomedical Engineering uses additional labels for devices serviced internally. This provides secondary notification of the device's status and PM compliance. The PM status of all devices is maintained within the web-based biomedical database and is available to all hospital staff.

Incoming Performance Verification

This verification is performed by the equipment manufacturer, a qualified third-party service provider, or by Biomedical Engineering. All devices are required to have a service record in the database indicating incoming performance verification and be appropriately labeled.

Special Devices and Systems

In the case of systems and devices that are beyond the expertise of Biomedical Engineering, such as anesthesia systems, X-ray systems, and lasers, it is recommended that these devices be performance tested by the manufacturer at the time of delivery or by a third party prior to deployment and use. The service provider is required to complete a field service report (FSR) or biomedical service report indicating that an incoming performance test was performed. Incoming performance testing documentation must be entered into the database. Upon entry of the incoming performance testing documentation, an inventory label and inspection sticker will be provided by Biomedical Engineering and attached to the device.

Biomedical Engineering Supported Devices and Systems

In cases where the incoming performance testing is performed directly by Biomedical Engineering staff, all inventory, labeling, incoming performance testing, and supporting documentation are automatically entered into the database and appropriate labels are attached to the equipment by Biomedical Engineering.

Repair Maintenance Service Request and Modalities

Various methods are used to repair, calibrate, and perform preventative maintenance on equipment at Intelligent Hospital. These methods include:

- Internal service by Biomedical Engineering
- Time and Material Services purchased from manufacturers and/or third-party sources
- Service contract

The decision of service mode is based on discussions between the clinical equipment user, Hospital Administration, Biomedical Engineering, and Finance. The service requirements, specifically preventive maintenance, are defined through Biomedical Engineering utilizing the standards established by the Commission on Medicare and Medicaid, TJC, ECRI, and/or manufacturer recommendations, consistent with the policies and standards identified in this document.

Devices Serviced by Biomedical Engineering

Biomedical Engineering has written procedures to reference when medical equipment fails, including using emergency clinical interventions and backup equipment.

For device service, requests can be provided directly through the web-based database application, or via direct service call to Clinical Engineering, or through the Center Help desk. On-call technicians will provide emergency service after hours.

Devices Serviced on Time and Materials Basis: Insurance or Service Contract

For devices not serviced directly by Biomedical Engineering staff, all service requests, including preventative maintenance, are generated by the user directly to the service provider of choice. It is the responsibility of the user to ensure that preventive maintenance is performed in accordance with the PM interval and schedule identified within the database. The user is required to ensure that the service provider supplies a FSR indicating a detailed description of the services provided. All service documentation is to be entered into the database utilizing the "Service Documentation and Archival Process" outlined below. The user is required to initiate the service process, receive a descriptive report of the service, and enter the service documentation into the database using the identified process.

Biomedical Engineering oversees preventative maintenance compliance for these devices, reviewing device status monthly to ensure the robustness of the preventative program, and reports all devices with pending or overdue PM to the user departments, administration, and the Safety Committee. Biomedical Engineering works with the user departments to support the PM process and correctly document the device status.

Service Documentation and Archival

For devices in the MEMP, all device services, including failure-based repairs, preventative maintenance, recall remediation, and incident management are documented in the Biomedical Database associated with the uniquely identified devices.

Internally Supported Devices

For all device services performed by Biomedical Engineering, each service is categorized by a work order and is classified as follows:

- Incoming Acceptance Testing
- Failures (where possible and applicable including failure mode)
  - PM/PV Preventable Service

- Degree of Equipment Failure
  - Major
  - Minor
- Type of Equipment Failures
  - Spontaneous
  - User or Process Related (Previously Defined as User Error)
  - Wear and Tear Related

- Recall
- Incident
- Preventative Maintenance (PM)/Performance Verification (PV)
- Upgrade

Devices Supported through Third Party, Service Contract, Warrantee, etc.

All service providers, whether the device is under warranty, service contract, or serviced on a time and materials basis, are required to provide a FSR to the user and/or Biomedical Engineering at the time of service. The user shall identify the FSR with the unique identifier (barcode ID#) and identify whether the service was a repair or preventive maintenance and provide the FSR/report to Biomedical Engineering for processing and database entry.

Risk Assessment

Biomedical Engineering manages medical equipment risks. Biomedical Engineering identifies critical (high-risk) medical equipment in the inventory for which there is a risk of serious injury or death to a patient or staff member should the equipment fail. All critical (high-risk) equipment that is in the inventory is identified as such.

Critical (High-Risk) devices are those that in the case of failure have a strong potential for causing serious injury and/or death to the patient or staff. All devices defined as "***Life Support***" are defined as critical (high-risk) devices. **All critical (high-risk) devices are required to maintain a PM/PV compliance of 100%** [10]**.**

Not Critical (Not High-Risk) devices are those that in the case of failure do not cause direct injury to the patient. **Not critical (not high-risk) devices are required to maintain a PM/PV compliance of 95%.**

Preventative Maintenance/Performance Verification (PM/PV)

Biomedical Engineering operates an adaptive preventive maintenance/ performance verification program to continuously monitor and evaluate the operational integrity and safety of the medical devices. The PM/PV process ensures increased equipment reliability, patient safety, and regulatory compliance. Biomedical Engineering evaluates systems/devices, to establish the appropriate preventive maintenance criteria, interval, and test specification, based on their inclusion into specific programs: OEM, AEM, and general programs.

The Intelligent Hospital manages all preventative maintenance according to the manufacturer's recommendations and guidelines. Any exceptions are presented and approved by the Institution's Executive Safety Committee.

PM Intervals

The PM intervals are established by Biomedical Engineering based on guidelines of CMS, ECRI Institute, manufacturer recommendations, good engineering practices, and service experiences within MSKCC. For devices within the OEM program, all PM/PV intervals and test procedures are specifically defined by the manufacturer. In all other programs, Biomedical Engineering can assess the requirements for PM and, based on the hospital's experience, can alter the interval and/or the procedure (AEM Program). This requires the presentation of supporting information and approval by the Executive Safety Committee.

PM Compliance Monitoring

The Biomed Database automatically monitors and tracks the PM status of all devices included in the MEMP. A device is identified as PM due once the PM cycle interval is exceeded. PM overdue notification is dynamically identified, and notification is sent to the user department for all devices in the MEMP that have exceeded 30 days beyond the PM due date.

Notification of PM Due

The database automatically determines and tracks the preventive maintenance (PM) status of all devices and identifies devices pending and overdue for PM. For devices serviced by Biomedical Engineering, a PM report is generated and distributed to the Biomedical Engineering staff to complete scheduled PM.

For devices serviced externally, on a bimonthly basis, for devices not serviced directly by Biomedical Engineering, the PM overdue information is e-mailed directly to the assigned user contacts, requesting that PM be performed or the appropriate and supporting documentation be provided or directly entered into the database. The database dynamically maintains the status of all devices, which can be accessed online through the MSK intranet.

PM Overdue Escalation Process

To ensure PM compliance throughout the institution, the following notification and escalation process has been established. Follow-up notification of the PM status and non-compliance will be issued to the department administrators and associate administrators as follows:

- PM Overdue Notice Issued Bimonthly → Department Administrator
- Six-Month Overdue Notice → Hospital Administration

Equipment Removal (Retired) from the Program

Biomedical Engineering reviews the status of equipment based on safety, availability of parts or repair services, failure history, and life expectancy. If equipment can no longer be properly or safely supported, the equipment is presented to the User Departments/owners and Product Evaluation and Safety Committees and identified for replacement planning and scheduled removal.

Equipment is retired from service by the user department and is to be removed from the active database and all labeling on the equipment, including the barcode ID #, is to be removed. Devices are retired, only when they are removed from the facility, discarded, sold, donated, or provided to biomedical engineering to be used for spare parts. Devices not in use or placed in storage for future use are to be maintained within the database for inventory purposes, and any required preventive maintenance is discontinued. This equipment will be identified as Inactive in the database, and all device and associated historical information included as part of the

MEMP, whether current, inactive, or retired, are retained within the database for a period not less than 10 years.

Serviced by Biomedical Engineering

Biomedical Engineering attempts to locate devices requiring PM during the due month and the following three consecutive months, working directly and in coordination with the user departments and staff. At the end of the fourth month, notification is provided to the user department of pending device inactivation. Devices overdue for PM for greater than 4 months are automatically defined as inactive within the active inventory and the user/owner of the equipment is notified.

Devices Serviced by OEM or Third Party Service Providers

Whether the device is serviced on a time and materials basis or service contract**, it is the responsibility of the user to ensure that all PM compliance is current and reflected in the database.** At the end of a year, PM overdue cycle (devices overdue for PM for greater than or equal to 1 year) devices non-compliant will be automatically defined as inactive and the user department is notified.

Cybersecurity

The institution will adopt a cybersecurity program to include the features and capabilities:

1. Identify all medical devices networked (wired or wireless) susceptible to cybersecurity penetration and attack.
2. Each identified device will be assessed to define the magnitude of risk it represents in terms of cyberattack.
3. Medical Systems, including medical devices, transducers, integration and connectivity with clinical software applications/systems, network infrastructure, and servers, should identified to identify broader-based potential vulnerability.
4. Medical devices will be associated with controls to minimize vulnerabilities, based on operational and life cycle considerations.
5. Medical devices will be monitored to dynamically identify potential threat activity.
6. Establish a process to respond to specific, prioritized threat activities.

Recall and Hazard Alerts

Equipment incidents, recalls, and hazard alerts are managed in accordance with existing MSK administrative, safety, and biomedical engineering policies and procedures. MSK subscribes to several organizations (i.e., ECRI) providing Device Alert notification of device recalls and hazard alerts. In addition, the FDA and OEM notify the owner of these devices of pending issues and recalls.

In cases where the devices are serviced by biomedical engineering staff, all services are coordinated with the user department and directly entered into the database. Devices under service contract, T&M, warranty, or third-party support are the responsibility of the service provider and managed by Biomedical Engineering. All service from these organizations is coordinated with the department and documented in the database.

Incidents

MSK departments are required to notify Biomedical Engineering in the event of a device-related incident causing injury to either a patient or staff. Biomedical Engineering provides a rapid response to these conditions and tests and evaluates the incident in accordance with Biomedical Engineering expertise to assess system status. Based on the Biomedical Engineering assessment, the device is sequestered and Biomedical Engineering works with Risk Management to determine the appropriate actions, which can include testing by a third party and/or OEM.

In addition, Biomedical Engineering routinely monitors all reported incidents within the Center's MSK RISQ system for device-related incidents. Where possible, Biomedical Engineering identifies the device involved in the incident, verifies device operation and integrity, and sequesters the device, if necessary, for further evaluation.

Annual Review

On an annual basis, Biomedical Engineering, in conjunction with the Executive Safety committee, evaluates the performance of the MEMP to identify enhancements, new technologies, or changes in policy to ensure regulatory compliance and ensure overall patient safety from injury associated with medical device failure or misuse.

References

1. Department of Health & Human Services, Centers for Medicare & Medicaid Services Critical Access Hospital (CAH) Equipment Maintenance Requirements Center for Clinical Standards and Quality/Survey & Certification Group, Ref: S&C: 14:41-CAH, dated August 8, 2014. http://www.cms.gov/Medicare/Provider-Enrollment-and-Certification/SurveyCertificationGenInfo/Downloads/Survey-and-Cert-Letter-14-41.pdf
2. The Joint Commission Revised Equipment Maintenance Requirements, Environment of Care (EC), effective January 1, 2015. http://www.jointcommission.org/assets/1/18/Revised_Equipment_Maintenance_Standards_for_Critical_Access_Hospitals.pdf
3. ECRI Home Page, https://www.ecri.org/Pages/default.aspx
4. FDA Medical Devices section, http://www.fda.gov/MedicalDevices/
5. CMS (from 1 above) Section D: When AEM is not permissible.
6. CMS (from 1 above) Section B and C: Decision to place equipment in an AEM.
7. http://www.mdsr.ecri.org/summary/detail.aspx?doc_id=8280
8. http://www.mdsr.ecri.org/summary/detail.aspx?doc_id=8280
9. https://one.mskcc.org/sites/pub/ha/Policies/HA-9106.pdf
10. Page 2 Bottom Section. http://www.jointcommission.org/assets/1/23/jconline_October_29_14.pdf

Appendix 2: Medical Systems – Device Inventory Management: Medical Device Inventory Management System Data Content

Medical Device Description

☐ Inventory Number:	Unique as set identifier in the inventory database
☐ Equipment Category (list):	Medical/Non-Medical/Research
☐ Equipment Type:	Naming Convention for the classification of devices ECRI compliant for medical devices
☐ Item Alias:	Alternate Name for grouping classification (Pumps)
☐ Alternate Item Name:	Any other asset description used by different departments to identify asset (ESU, FT10, Medtronic)
☐ Serial Number:	A unique manufacturer identifier
☐ Model Number:	Manufacturer-issued combination of numbers and letters used to identify a mass-produced product

☐ MFR Device Name:	Manufacturer Name of Product
☐ Manufacturer:	A company that produces equipment in large quantities ECRI complaint for the medical equipment category
☐ Critical Equipment:	Devices identified the critical device category High risk equipment that could potentially lead to severe complications or death because of malfunction
☐ Life Support:	Subsystems defined as Critical equipment A device that supplants a vital integral organ function
☐ Equipment Status:	Active, Discarded, Out for Repair…
☐ Equipment Usage:	Clinical, Research, Office, Other…
☐ Medical Device Class:	FDA definition Class 1, Class 2, Class 3
☐ Comments:	General comments field
☐ Alias Inventory Number:	Other numbers used to identify devices
☐ Security System:	Network [Y/N] Subject to Cyber vulnerability
☐ Software Version:	Operating Firmware or Software
☐ Inspection Work Order:	Displays pending work orders that require attention

Equipment Type and Usage

- ☐ Clinical
- ☐ Medical Furniture
- ☐ Surgical Instrument
- ☐ Office
- ☐ Research
- ☐ Other

Medical Equipment Management Program Parameters

☐ MEMP:	Included in the Medical Equipment Management Program (Yes/No)
☐ Service Policy:	Two options
☐ OEM:	Equipment/devices which are managed according to the original equipment manufacturer's recommendations for service

☐ AEM:	All eligible equipment which can be maintained using alternate support strategies as defined by qualified staff. Subject to review by the Administrative Safety Committee or Legal.
☐ CMS Managed Category:	Medical Equipment defined as part of the MEMP program, which are mandated to follow OEM procedures and maintain manufacturer PM intervals.
☐ Imaging/Radiologic	
☐ Medical Laser	
☐ New to market Equipment	

Financial Data

The section describes how service and repair are paid.

☐ Service Provider or Manager	
☐ Biomed	
☐ Grant Funded	
☐ Insurance	
☐ Service Contract	
☐ Service on Demand	
☐ Cost Center:	Department Identifier, ownership
☐ Purchase Order Number:	Service purchase order number
☐ Next Year Disposition:	Service changes effective from the next year
☐ Annual Service Budget:	Estimate Service Cost/Expense
☐ Service Contract Expiration:	Service contract end date
☐ Service Contract Number:	Internal service contract number
☐ Service Cost to Date:	Total service cost from asset purchase date, Annual, other
☐ Financial Cost Center:	Used for billing financial system
☐ Fund Number:	External Source of Funding (Grants, etc.)

Service Contract Details

☐ Manufacturer:	A company that produces equipment (OEM) or other contracted service provider
☐ Service Status:	Displays the current service status for the asset
☐ Service Start Date:	On: Service contract start date
☐ Service Expired End:	Service contract end date
☐ Equipment Type:	Nomenclature used for the classification of devices ECRI compliant for medical devices.
☐ Model:	A combination of model number and MFR Device Name
☐ Serial Number:	Device unique identifier
☐ Ownership:	Division, Cost Centre, Department
☐ Email Contact	
☐ Notification	

Cybersecurity

- ☐ Main System Name
- ☐ Sub-System Name
- ☐ System Description
- ☐ Business Owner
- ☐ Technical Owner
- ☐ Risk Score
- ☐ Assigned To
- ☐ System Status
- ☐ Controls Implemented (Y/N)
- ☐ Controls Implementation Defined
- ☐ Main Security System Name
- ☐ Sub-Security System Name
- ☐ Assessment Name
- ☐ Assessment Description
- ☐ Assessment Type

Capital Replacement Planning

☐ Inventory Number:	Unique asset identifier in inventory database
☐ Capital Replacement	
☐ Device Category	
☐ General Medical	

- ☐ Lab Medicine/Path
- ☐ Periop/Anesth
- ☐ Pharmacy
- ☐ Radiation Oncology/ Medical Physics
- ☐ Radiology
- ☐ CRC Equipment Category: General device group that's based on department and usage

☐ Useful Life:	Estimated number of years in which a device will remain in service
☐ Estimated Replacement Cost	
☐ Purchase Cost:	MSRP for new equipment
☐ Recommended Year:	Recommended year of replacement after all factors are considered
☐ Projected Target Year:	Target date of equipment replacement

RTLS Location

☐ RFID Tag MAC Address:	Unique identifier for RTLS tags
☐ Original Location:	Last reported location for an asset
☐ Building	
☐ Floor	
☐ Room	
☐ Real-Time location:	Automatically updated location from RTLS system

Preventative Maintenance

- ☐ PM Due – Preventive maintenance due date
- ☐ PM Overdue – Items that have not been serviced by the scheduled PM due date. The last PM due date will be displayed.
- ☐ PM Priority – Low, medium, high
- ☐ PM Interval – (list) – PM interval frequency

- ☐ PM Service Provider – (list) – Indicates who is responsible for providing periodic maintenance
- ☐ Last PM Inspection (auto) – Populated from Service History by actual date of service
- ☐ Inspections – Any additional inspections required by asset (Laser Safety, QA)
- ☐ Inspection Type
- ☐ Inspection Interval
- ☐ Inspection Provider
- ☐ Last Inspected
- ☐ Next Inspection due

Network Specifications

The section is used for assets which has an attached software component.

- ☐ Software Serial Number
- ☐ IX Network Mask
- ☐ IX Network IP
- ☐ MAC Address
- ☐ MC Network IP
- ☐ MC Network Mask
- ☐ Firmware Version
- ☐ Software Version
- ☐ License Type
- ☐ License Quantity
- ☐ Leakage
- ☐ Workstation ID
- ☐ Server Name
- ☐ Wi-Fi SSID
- ☐ Wi-Fi Security Type
- ☐ Wi-Fi Network Segment
- ☐ MUSE Port #
- ☐ MUSE Location Code

Software Classification

- ☐ Medical Device Type
- ☐ MDDS
- ☐ Software
- ☐ Non-MDDS system
- ☐ Medical Device Class (OEM/FDA Classification)

- ☐ Class 1
- ☐ Class 2
- ☐ Class 3
- ☐ Application Owner (s) – Who owns the system
- ☐ Support Contact Information (s) – Who supports the system

Life Cycle Information

This section contains information about asset history (sequence of events).

☐ Date Purchase	
☐ Deactivation Date:	Displays the date an item was removed from service
☐ Date Inventoried:	Initial inventory date
☐ Last Serviced Date	
☐ Last Safety Inspection	
☐ Warranty Starts On:	Displays the first date of warranty coverage
☐ Warranty Expires On:	Displays the last date of warranty coverage
☐ Equipment Receipt Date:	Date of receipt at MSKCC
☐ Installation Completion Date:	The vendor installation completion date
☐ Date Entered into the Service:	The first date of usage
☐ System End of Life:	Manufacturer issued end of life date
☐ System End of Support:	The last date of support from the manufacturer
☐ Estimated Life:	The useful life of an asset. Defined by the manufacturer

Contact Information

This section contains main contact information for the asset.

- ☐ Department Contact Name
- ☐ Department Contact Phone
- ☐ Department Contact ID (LDAP)
- ☐ PM Service Provider Contact

Laser Safety Program

☐ Laser Safety Program:	Yes/No
☐ Laser Class:	An option to select the class of a medical laser
☐ Laser Wavelength:	The distance between two corresponding peaks of a wave
☐ Laser Power:	Laser output power
☐ Last Safety Inspection:	Populated based on laser safety inspections

Equipment Status

☐ In Clinical Service:	Active equipment in use for clinical service; two statuses
☐ Active	
☐ Reserve/Use as Backup	
☐ Removed from Clinical Service:	Devices that were temporarily removed from service
☐ Out for Repair	
☐ Not Used at Present Time	
☐ Out of Service/Not Found	
☐ Pending Security Approval	
☐ Inactive:	Equipment that is no longer used for clinical service and is no longer on-site
☐ Long-term Storage	
☐ Could Not Be Located After an Extended Period of Time	
☐ Retired:	Equipment that is permanently removed from service; six statuses
☐ Discarded	
☐ Removed by Vendor	
☐ Removed by research/project Principal Investigator	

- ☐ Cost Centre Closed
- ☐ Donated
- ☐ Sold

Service Responsibility

- ☐ Biomedical Engineering In-house Service
  - ○ General Patient Care Device
  - ○ Radiological/Imaging
  - ○ High Energy (Linear Accelerator)
  - ○ Laboratory Medicine
  - ○ Research Laboratories
    - ▪ Service Contract
    - ▪ Warrantee
    - ▪ Third Party
- ☐ IT
- ☐ Facilities
- ☐ User Defined

Cybersecurity

- ☐ Main System Name
- ☐ Sub-System Name
- ☐ System Description
- ☐ Businccs Owncr
- ☐ Technical Owner
- ☐ System Description
- ☐ Risk Score
- ☐ Controls Implemented (Y/N)
- ☐ Controls Implementation Notes
- ☐ Assessment Name
- ☐ Assessment Description
- ☐ Assessment Type
- ☐ Assessment Status

Appendix 3: Glossary of Terms

3D Printing 3D printing, or additive manufacturing, is a way of making objects by depositing layers of material based on a digital model.

802.11 Standard A wireless local area network (WLAN) standard in the 2.4, 3.6, and 5 GHz frequency bands. It is maintained by the IEEE LAN/MAN Standards Committee (IEEE 802). 802.11 Standard is the basis for the Wi-Fi specification. This technology is used by some RTLS solutions to calculate location.

802.15.4 Standard A standard which specifies the physical layer and media access control for low-rate wireless personal area networks (LR-WPANs). It is maintained by the IEEE 802.15 working group. 802.15.4 Standard is the basis for the ZigBee specification. This technology is also used by some RTLS solutions to calculate location.

Accountable Care A healthcare delivery and payment model that ties provider reimbursement to quality metrics and reductions in total cost of care. The Accountable Care Organization (ACO) is a group of coordinated healthcare providers that provides care to a given population.

Active RFID Systems In active RFID systems, tags have their own transmitter and power source, usually a battery. Active tags broadcast their own signal to transmit the information stored on their microchips. Active RFID systems typically operate in the UHF band and offer a range of up to 100 m. In general, active tags are used on large objects, such as rail cars, big reusable containers, and other assets that need to be tracked over long distances. They are more expensive than passive tags.

Additive Manufacturing Additive manufacturing, or 3D Printing, is a way of making objects by **depositing layers of material** based on a digital model.

Artificial Intelligence The ability of a digital computer or computer-controlled system to perform tasks commonly associated with intelligent beings. The term is frequently applied to the project of developing systems endowed with the intellectual processes characteristic of humans, such as the ability to reason, discover meaning, generalize, or learn from experience.

Application Software that utilizes data coming from middleware and that directly interacts with the end user. Examples of applications include asset tracking and management, patient flow, temperature monitoring, infection control and hand hygiene, staff duress, inventory tracking and management, positive patient identification, business intelligence and reporting, and wireless nurse call.

Air Interface Protocol A radio-based communication link protocol that governs how tags and readers communicate.

Augmented Reality Augmented reality (AR) is an interactive experience that combines the real world and computer-generated content. The content can span multiple sensory modalities, including visual, auditory, haptic, somatosensory, and olfactory. AR can be defined as a system that incorporates three basic features: a combination of real and virtual worlds, real-time interaction, and accurate 3D registration of virtual and real objects.

Backscatter A reflection of waves back in the direction from which they came. RFID tags use backscatter technology to reflect radio waves back to the reader, usually at the same carrier frequency. The reflected signal is modulated to transmit data.

Bandwidth The range or band of frequencies, defined within the electromagnetic spectrum, that a system is capable of receiving or delivering.

Barcodes Consist of small images of lines (bars) and spaces affixed to retail store items, ID cards, and postal mail to identify a particular product number, person, or location. A barcode reader uses a laser beam that is sensitive to the reflections from the line and space thickness and variation. The reader translates information from the image to digital data and sends it to a computer for storage or for another process. 2D barcodes store information not only horizontally, as one-dimensional barcodes do, but vertically as well. That construction enables 2D codes to store up to 7,089 characters. The traditional, uni-dimensional barcode has only a 20-character capacity.

Battery-Assisted Passive Tag (BAP) RFID tags, with batteries, that communicate using the same backscatter technique used by passive tags (tags with no batteries). They use the battery to run the circuitry on the microchip and sometimes an onboard sensor. They have a longer read range than a regular passive tag because all of the energy gathered from the reader can be reflected to it. They are sometimes called "semi-passive RFID tags."

Biometrics Technology used to identify individuals by comparing biological data, such as fingerprints, voice characteristics, and iris patterns, against stored data for that individual. Biometric systems consist of a reader or scanning device, software that converts the scanned biological data into a digital format and compares match points, and a database that stores the biometric data for comparison. Authentication by biometric verification is becoming increasingly common in corporate and public security systems, consumer electronics, and point of sale (POS) applications. Specific biometric AIDC (Automatic Identification and Data Capture) technologies include finger-scanning, electro-optical fingerprint recognition, finger vein ID, and voice recognition.

Bioprinting Bioprinting (also known as 3D bioprinting) is a combination of 3D printing with biomaterials to replicate parts that imitate natural tissues, bones, and blood vessels in the body.

Biotechnology Biotechnology, the use of Biology to solve problems and make useful products. The most prominent area of biotechnology is the production of therapeutic proteins and other drugs through genetic engineering.

Bluetooth Low Energy (BLE) A wireless network protocol, especially suited for sensors and other small devices, that requires extremely low power consumption. The protocol enables devices such as an iPhone to transmit a Bluetooth signal to beacons, with a read range up to 50 m.

Connectivity A generic term for connecting devices to each other in order to transfer data back and forth. It often refers to network connections, which embrace bridges, routers, switches, and gateways as well as backbone networks.

Coronavirus (COVID-19) Coronavirus disease 2019 is a contagious disease caused by severe acute respiratory syndrome coronavirus 2. The first case was identified in Wuhan, China, in December 2019. Symptoms of COVID-19 are variable, but often include fever, cough, fatigue, breathing difficulties, and loss of smell and taste.

Cybersecurity Cybersecurity is the practice of protecting critical systems and sensitive information from digital attacks. Also known as information technology (IT) security, cybersecurity measures are designed to combat threats against networked systems and applications, whether those threats originate from inside or outside of an organization.

Data Retention The ability of a microchip to maintain the information stored in EEPROM (Electrically Erasable Programmable Read-Only Memory). RFID tags and other microchips can typically retain data for 10 years or more, but data retention depends on temperature, humidity, and other factors.

Environment of Care The Environment of Care is made up of three basic elements: for each specific patient care environment (surgical, radiation, critical care, imaging, laboratory, etc. the specific building or space requirements, including how it is arranged and the special features that protect patients, visitors, and staff. Equipment used to support patient care or to safely operate the building or space.

E-pedigree or Electronic Pedigree An electronic document that provides data related to the history of a batch of a drug. The e-pedigree is used to identify a drug prior to the sale, purchase, or trade of that product. The states of California and Florida have set deadlines for mandatory compliance with an e-pedigree system for tracking of medications through the supply chain. The FDA has also established electronic pedigree regulations to reduce the risk of counterfeit products.

Electromagnetic Interference (EMI) A phenomenon in which the electromagnetic field of one device disrupts, impedes, or degrades the electromagnetic field of another device, potentially blocking transmission.

Electromagnetic Spectrum The range or continuum of electromagnetic radiation, characterized in terms of frequency or wavelength.

Electronic Product Code (EPC) A unique identifier for every physical object anywhere in the world, for all time. Its structure is defined in the GS1 EPCglobal Tag Data Standard; an open standard available for download from the GS1 EPCglobal website. The EPC is designed as a flexible framework that can support a variety of existing coding schemes, including many coding schemes currently in use with barcode technology. EPC identifiers currently support seven identification keys from the GS1 system of identifiers. An EPC takes the form of a 96-bit string of data stored in a RAIN RFID tag chip that can act as a key to a database. The EPC is a Uniform Resource Number (URN) namespace registered to EPCglobal by the Internet Assigned Numbers Authority (IANA).

Encryption Translation of data into a code for the purpose of keeping information secure from all but the intended recipient.

EPC global A joint venture between GS1 (formerly known as EAN International) and GS1 US. It is an organization set up to achieve worldwide adoption and standardization of Electronic Product Code (EPC) technology.

Excite The transmission of radio frequency energy from the RFID tag reader to stimulate a passive RFID tag to provide power to transmit its data back.

Frequency The number of cycles a periodic signal executes in unit time. It is usually expressed in Hertz (cycles per second) or appropriate weighted units such as kilohertz (kHz), megahertz (MHz), and gigahertz (GHz).

GS1 An international organization to develop and maintain standards for supply and demand chains. The four key standards of focus are: Barcodes (used to automatically identify things), eCom (electronic business messaging standards allowing automatic electronic transmission of data), GDSN (Global Data Synchronization standards that allow business partners to have consistent item data in their systems at the same time), and EPCglobal (which uses RFID technology to immediately track an item).

Health Insurance Portability and Accountability Act (HIPAA) The US Congress enacted HIPAA in 1996 to regulate the interchange of private patient data to help prevent unlawful disclosure or release of medical information.

Healthcare Information and Management Systems Society (HIMSS) It is a not-for-profit association of 50,000 individual members and 570 corporate members. This organization is dedicated to improving healthcare quality, safety, access, and cost-effectiveness through the use of information technology and management systems.

Healthcare Information Exchange (HIE) The process by which health information is mobilized electronically between or across organizations within a region, community, or hospital system. Federal and state regulations regarding HIEs are still being defined. In the meantime, multiple state and healthcare provider exchanges have been developed in the United States to manage the movement of electronic records.

High-Frequency (HF) The frequency bandwidth from 3 to 30 MHz HF RFID tags typically operate at 13.56 MHz, can normally be read at short range (three feet or less), and transmit data faster than low-frequency tags, although they consume more power than low-frequency tags.

Infrared (IR) A technology that uses electromagnetic radiation with a wavelength that is longer than that of visible light, but shorter than that of microwaves and terahertz radiation. The IR signal does not penetrate walls, ceilings, floors, or large objects inside a room, but it does bounce off any object in its path. This technology is used to enable RTLS systems and is often used in conjunction with RFID.

Injectable Tech Technology that is placed inside the body using an instrument to pierce the skin or mucous membranes.

Insertable Tech Technology that is inserted in a natural orifice of the body (ear canal, tear duct, oral cavity, etc.).

Intelligent Hospital The intelligent hospital is one that works better and smarter. It's better because it's resourceful, creative, and perceptive about what patients and doctors need, and it's smarter because it's astute and inventive when it comes to weaving together diverse technologies to enhance patient care.

Intensive Care Unit (ICU) A department within a hospital or healthcare facility dedicated to intensive care of patients with severe or life-threatening injuries or illnesses that require constant, close monitoring and support.

Interactive Voice Response (IVR) A telephony technology in which an individual uses a touch-tone phone to interact with a database to acquire information from, or enter data into, the database.

International Organization for Standardization (ISO) A non-governmental organization consisting of the national standards institutes of 205 countries. Each member country has one representative, and the organization maintains a Central Secretariat in Geneva, Switzerland, that coordinates the system. Most RFID related ISO Standards are: ISO 10536 – the international standard for proximity cards; ISO 11784 – the international standard defining frequencies, baud rate, bit coding, and data structures of the transponders used for animal identification; ISO 14443 – a set of international standards covering proximity smart cards; ISO 15693 – the international standard for vicinity smart cards; ISO 18000 – a series of international standards for the air interface protocol used in RFID systems for tagging goods within the supply chain; ISO 7816 – a set of international standards covering the basic characteristics of smart cards, such as physical and electrical characteristics, communication protocols, and others.

Interoperability Interoperability is a characteristic of a product or system to work with other products or systems. While the term was initially defined for information technology or systems engineering services to allow for information exchange,

Joint Commission on Accreditation of Healthcare Organization (JCAHO) A non-profit, US association that accredits healthcare organizations and programs as part of its mission to improve healthcare. A majority of state governments recognize the Joint Commission's accreditation as a condition of license and the receipt of Medicaid reimbursements.

LAN (Local Area Network) A relatively small network covering areas such as a room, a department, a building, a campus, etc.

Low-Frequency (LF) The frequency bandwidth from 30 to 300 kHz. Low-frequency RFID tags typically operate at 125 or 134 kHz. Low-frequency RFID tags must be read from within three feet, and their data transfer rate is slow, but they are less susceptible to interference than UHF tags.

Machine to Machine (M2M) The term for technologies that enable devices (such as sensors or meters) to communicate with each other or another device (such as an appliance). An M2M network can consist of a connection between two devices or multiple devices. M2M healthcare applications include patient monitoring solutions or drug dispensing tracking.

Memory A means of storing data in electronic form. A variety of random access (RAM), read-only (ROM), Write Once-Read Many (WORM), and read/write (RW) memory devices can be used. In RFID terms, memory is the amount of data that can be stored on the microchip in an RFID tag. It can range from 64 bits to 2 kilobytes or more on passive tags.

Mesh Network A Mesh network consists of wireless nodes that relay data to other nodes as part of a network that ultimately forwards data to a server. An example of a mesh network is Zigbee digital radio technology that forwards data via "hops" from one node to another.

Micro-Electrical Mechanical Systems (MEMS) A term that refers to the combining of electrical and mechanical components on a chip to produce a very small system. MEMS medical devices are being developed to replace traditional devices because they are so small. The MEMS devices promise to be more sensitive and robust than traditional technology. MEMS technology can be used within a host body or in biological samples to detect and diagnose health status.

Microwave Tags A term that is sometimes used to refer to RFID tags that operate at 5.8 GHz. They have very high transfer rates and can be read from as far as 30 feet away, but they use a lot of power and are expensive. (Some people refer to any tag that operates above about 415 MHz as a microwave tag.)

Middleware In the RFID world, this term refers to software that resides on a server between readers and enterprise applications. The middleware is used to filter data and pass on only useful information to enterprise applications. Some middleware can also be used to manage readers on a network.

Modulation A term to denote the process of superimposing (modulating) channel encoded data or signals onto a radio frequency carrier to enable the data to be effectively coupled or propagated across an air interface. Modulation is also used as an associative term for methods used to modulate carrier waves. Methods generally rely on the variation of key parameter values of amplitude, frequency, or phase. Digital modulation methods principally feature amplitude shift keying (ASK), frequency shift keying (FSK), phase shift keying (PSK), or variants.

Near Field Communication (NFC) A set of standards for smartphones and tablets to establish radio communication with each other or tags by touching them or bringing them into close proximity, usually no more than a few centimeters from each other or a tag. Existing and anticipated applications include contactless transactions, data exchange, and simplified setup of more complex communications such as Wi-Fi.

Outcomes Measures Used to assess the impact of health services in terms of improved quality and/or longevity of life and function.

Passive RFID Systems In passive RFID systems, the reader sends a radio signal to the tag. The RFID tag then uses the transmitted signal to power on and reflect energy back to the reader. Passive RFID systems can operate in the LF, HF, or UHF radio bands. Passive tags do not require a power source or transmitter and only require a tag chip and antenna. They are cheaper, smaller, and easier to manufacture than active tags and typically have a range of less than 15 m.

Patient Activation/Engagement The extent to which patients have skills, knowledge, and motivation to participate as part of their care team.

Patient Centric Identification and Association Development of a standardized naming or identification methodology to enable the identification of a patient, staff, medical devices, supplies, and pharmaceuticals associated with a specific patient. This ensures specific processes commonly associated with patient care and delivery are uniquely identified and associated for process management and workflows. As an example, alarm management and delivery (ensure delivery of alarms to the appropriate clinical staff), sample collection (associate blood samples, etc. to the correct patient), and medication administration (ensure delivery of the right drug to the right patient at the right time). This identification and association process minimizes errors and enhances patient safety.

Population Management The health outcomes of a group of individuals, as well as the distribution of such outcomes within the group; an approach that aims to improve the health of an entire human population.

Protocol A set of rules governing a particular function, such as the flow of data or information in a communication system.

Radio Frequency Identification (RFID) RFID is a form of wireless communication that uses radio waves to identify and track items. RFID can tell you what an object is, where it is, and even its condition, which is why it is integral to the development of the Internet of Things – a globally interconnected web of objects allowing the physical world itself to become an information system. RFID systems can be broken down by the frequency band within which they operate: low frequency (LF), high frequency (HF), and ultra-high frequency (UHF). There are also two broad categories of RFID systems.
See Active RFID Systems and Passive RFID Systems.

RAIN RFID RAIN RFID is a passive wireless technology system that connects billions of everyday items to the Internet, enabling businesses and consumers to identify, locate, authenticate, and engage each item. RAIN RFID is used in a wide variety of applications, including inventory management, patient safety, asset tracking, and item authentication. RAIN is the fastest growing segment of the RFID market and uses a single, global standard: UHF Gen 2 (ISO/IEC 18000-63). The RAIN RFID industry is represented by a global alliance called the RAIN Alliance.

Read Only Term applied to a transponder or tag in which data is stored in an unchangeable manner and can therefore only be read and not altered. Writing to a read-only tag is also impossible.

Read Range The distance from which a reader can communicate with a tag. Active tags have a longer read range than passive tags because they use their own power source (usually a battery) to transmit signals to the reader. With passive tags, the read range is influenced by frequency, reader output power, antenna design, and method of powering up the tag. Low-frequency tags use inductive coupling, which requires the tag to be within a few feet of the reader.

Read Rate The maximum rate at which data can be communicated between transponder and reader/interrogator, usually expressed in bits per second (bps or bits.s-1).

Read/Write When applied to an RFID system, this term refers to the ability to both read data from a transponder and to change data (write data) using a suitable programming device. See RFID Reader.

Reader/Gateway (Connectivity) A RAIN RFID reader or gateway is a device that provides the connectivity between tag data from an endpoint and the enterprise system software that needs the information. A reader connects to external antenna(s) while a gateway integrates the reader and antenna(s) into a single device. Readers and gateways communicate with tags (or endpoints) that are within their field of operation, performing any number of tasks, including simple continuous inventorying, filtering (searching for tags that meet certain criteria), writing (or encoding) selected tags, etc.

Readmission Term that describes an admission to an acute care hospital within 30 days of discharge from an acute care hospital.

Remote Care or Virtual Care Synonymous with Telemedicine or Telehealth, this term refers to the use of telecommunication and information technologies to provide clinical care at a distance.

Resolution Determines how well a tagged person or item can be tracked to a specific location (e.g., resident room, ER bay).

RF-IR A hybrid technology that uses both radio waves and infrared for identification and tracking purposes. This technology is used to enable some RTLS systems.

RFIDba The International RFID Business Association (RFIDba) was founded in April 2004 as a not-for profit, educational, technology and frequency agnostic, trade association dedicated to serving the business needs of the end user community with vendor neutral information on RFID and RTLS technologies along with information on other associated, complimentary technologies.

RFID in Healthcare Consortium The RFID in Healthcare Consortium (RHCC) was founded on September 13, 2008, initially to address EMI issues with RFID and RTLS technologies but has since transformed itself into a full-fledged trade organization. The Consortium is a primary source for vendor neutral, educational programs; industry information; and other value-added services pertinent to RFID and RTLS technologies. The RHCC serves the needs of end user communities in the healthcare, assisted living, and nursing home industries.

RFID Receiver A device that listens for RF (radio frequency) signals and converts them into data packets that are available for further processing. It is used with active RFID tags, which continuously emit pre-programmed messages.

RFID Transponder (Tag) A microchip attached to an antenna that can be applied to an object. The tag picks up signals from, and sends signals to, an RFID reader. The tag may contain a unique serial number and may have other information such as a customer account number. Tags come in many forms, such as smart labels that have a barcode printed on them, or the tag can simply be mounted inside a carton or embedded in plastic. RFID tags can be active, passive, or semi-passive.

Real-Time Locating System (RTLS) Refers to technology that is used to locate and track people and items (such as assets, equipment, inventory) by associating a tag with each person or item. This term is commonly used in reference to "active," battery-powered locating technologies.

Robotic Surgery Robotic surgery, also called robot-assisted surgery, allows doctors to perform many types of complex procedures with more precision, flexibility, and control than is possible with conventional techniques. Robotic surgery is usually associated with minimally invasive surgery – procedures performed through tiny incisions. It is also sometimes used in certain traditional open surgical procedures. The most widely used clinical robotic surgical system includes a camera arm and mechanical arms with surgical instruments attached to them. The surgeon controls the arms while seated at a computer console near the operating table. The console gives the surgeon a high definition, magnified, 3D view of the surgical site. The surgeon leads other team members who assist during the operation.

Step-down Unit A hospital nursing area dedicated to the care of patients who have undergone surgery. As opposed to the ICU, the Step-down Unit serves patients who are stable but may require monitoring due to recent surgery.

Swallowed Tech Technology that is employed after swallowing.

Tag (Endpoint) A RAIN RFID tag consists of an integrated circuit (called an IC or chip) attached to an antenna that can be built into apparel hang tags, labels, wristbands, security tags, industrial asset tags, etc. The tag contains a unique identifier or EPC and, when placed on an item, creates an endpoint that communicates with readers to deliver the identity, location, and authenticity of items.

Telehealth Sometimes called telemedicine. The use of telecommunications and information technology to deliver health services and transmit health information to another location.

Telemetry The automatic transmission and measurement of data from remote sources by wire, radio, or other means.

Transponder An electronic transmitter/responder, commonly referred to as a tag.

Ubiquitous Sensor Network (USN) A network of sensors providing coverage of every single part of an area, such as a healthcare facility, however remote. Such networks are typically managed from a control center.

Ultra-High-Frequency (UHF) The frequency bandwidth from 300 MHz to 3 GHz. Typically, RFID tags that operate between 866 and 960 MHz send information faster and farther than high- and low-frequency (HF and LF) tags. However, UHF signals cannot pass through items with high water content. UHF tags generally consume more power than low-frequency tags.

Ultrasound A technology that uses a cyclic sound pressure with a frequency greater than the upper limit of human hearing. The production of ultrasound is used in Real-Time Location Systems (RTLS), as it can provide room level location accuracy because the sound does not penetrate walls.

Ultra-Wideband (UWB) Any radio technology used at a very low energy level for short-range, short-duration, high bandwidth communications. UWB technology pulses within a bandwidth exceeding the lesser of 500 MHz or 20 percent of the center frequency. The UWB pulses are received by sensors which determine a tag's location based on Time-Difference-of-Arrival (TDoA) and Angle of Arrival (AoA). This technology is used to enable accurate indoor positioning for RTLS systems.

Unique Device Identification (UDI) A system used to mark and identify medical devices within the healthcare supply chain. The FDA has released a rule that a unique number should be assigned by the device manufacturer to each version or model of a medical device, in both human readable format and AutoID format (such as barcode or RFID).

Unique Patient Identifier (UPI) An identification code used in the management of healthcare information and record-keeping. The UPI is used to identify and access patient care information as well as for medical record chart analysis, billing, and reimbursement. The Joint Commission on Accreditation of Healthcare Organization (JCAHO)'s Information Management Standards mandate that the unique patient identifier be part of a patient's medical records.

Video Conferencing Also known as video-teleconferencing, this term refers to the use of a set of telecommunication technologies that allow two or more locations to communicate by simultaneous two-way video and audio transmissions.

Virtual Care See Remote Care or Telehealth.

WAN (Wide Area Network) As its name suggests, this is a computer network that covers a far wider area than an LAN, such as cities, countries, continents, and the whole world. A WAN is formed by linking LANs together.

Wearable Tech Technology that is worn on the body in the form of clothing, straps, hanging around the neck, socks, underwear, etc.

Wi-Fi Refers to any system that uses the 802.11 Standard, which was developed by the Institute of Electrical and Electronics Engineers (IEEE). Wi-Fi networks operate in the 2.4 and 5 GHz radio bands, with some products that contain both bands (dual band). Wi-Fi is a very common wireless technology that is used to connect machines in a LAN. This technology is used by some RTLS systems for locating purposes.

Wireless Medical Telemetry Services (WMTS) The remote monitoring of a patient's physiological parameters including pulse and respiration rates using RFID or other medical telemetry devices.

Wireless Sensor Network (WSN) A network of spatially distributed autonomous sensors to cooperatively monitor physical or environmental conditions including motion, temperature, pressure, sound, or vibration.

A sensor network is typically equipped with a radio transceiver or other wireless communications device, a small micro-controller, and an energy source – usually a battery.

Write The RFID process of transferring data to a transponder (tag) from a reader, as well as storing the data on the transponder, which may also encompass the reading of data to verify the data content.

Write Once Read Many (WORM) A label distinguishing a transponder that can be partially or totally programmed once by the user and thereafter only read.

Write Rate The rate at which data is transferred to a transponder and stored within the memory of the device and verified. The rate is usually expressed as the average number of bits or bytes per second over which the complete transfer is performed.

ZigBee See 802.15.4 Standard.

Index

Note: Locators in *italics* represent figures, locators in **bold** represent tables.

O

P

R

S

For Product Safety Concerns and Information please contact our EU representative GPSR@taylorandfrancis.com Taylor & Francis Verlag GmbH, Kaufingerstraße 24, 80331 München, Germany